*The human mind, once stretched by a new idea,
never goes back to its original dimensions.*

—Oliver Wendell Holmes

CONTENTS
AT A GLANCE

LEGO® Spybotics™
Secret Agent Training Manual

Ralph Hempel

LEGO® Spybotics™ Secret Agent Training Manual

ISBN (pbk): 1-59059-091-0

Printed and bound in Canada 12345678910

Technical Reviewer: David Koudys

Editorial Directors: Dan Appleman, Gary Cornell, Jason Gilmore, Simon Hayes, Karen Watterson, John Zukowski

Managing Editor: Grace Wong

Development and Copy Editor: Nicole LeClerc

Proofreaders: Beth Christmas, Brendan Sanchez

Production Manager: Kari Brooks

Compositor: Diana Van Winkle, Van Winkle Design Group

Artist: Cara Brunk, Blue Mud Productions

Book Interior and Cover Designer: Kurt Krames

Manufacturing Manager: Tom Debolski

Marketing Manager: Stephanie Rodriguez

Distributed to the book trade in the United States by Springer-Verlag New York, Inc., 175 Fifth Avenue, New York, NY, 10010 and outside the United States by Springer-Verlag GmbH & Co. KG, Tiergartenstr. 17, 69112 Heidelberg, Germany.

In the United States, phone 1-800-SPRINGER, email orders@springer-ny.com, or visit http://www.springer-ny.com.

Outside the United States, fax +49 6221 345229, email orders@springer.de, or visit http://www.springer.de.

For information on translations, please contact Apress directly at 2560 9th Street, Suite 219, Berkeley, CA 94710.

Phone 510-549-5930, fax: 510-549-5939, email info@apress.com, or visit http://www.apress.com.

CONTENTS

LIST OF TABLES

ABOUT THE AUTHOR

 **RALPH HEMPEL** (BASc.EE, P.Eng) is an independent consultant in the field of embedded systems. He provides systems design services, training, and programming to clients across North America. His specialty is in deeply embedded microcontroller applications, which includes alarm systems, automotive controls, and consumer electronics.

Ralph also provides training and mentoring for software development teams that are new to embedded systems and need an in-depth review of the unique requirements of this type of programming. He uses the LEGO MINDSTORMS system to reinforce important concepts in a fun, hands-on introduction to embedded systems issues.

Ralph holds a degree in electrical engineering from the University of Waterloo and is a member of the Professional Engineers of Ontario. He lives in Owen Sound, Ontario, with his wife, Christine, and their three children, Eric, Owen, and Graham.

ABOUT THE TECHNICAL REVIEWER

DAVID KOUDYS has created with LEGO bricks since his earliest recollection. He received his first Technic set in 1979. When MINDSTORMS came out, Dave immediately purchased a few sets, and he has enjoyed creating and programming autonomous robots ever since. Dave thinks the Spybotics robots are a great source of fun on their own and an awesome extension to the MINDSTORMS system.

Dave currently resides close to Winona, Ontario, Canada, and currently works for a major school bus transportation company, where he maintains their computer systems. He is a member of rtlToronto, a group of LEGO enthusiasts in the southern Ontario area. In addition to building robots out of LEGO bricks, Dave's hobbies include bicycling, camping, canoeing, reading, SCUBA diving, submarine building, HTML design, woodworking, and playing video games.

ACKNOWLEDGMENTS

This is the fourth book I've been involved with that deals with LEGO products. My first three books are on the subject of the LEGO MINDSTORMS system that was released in late 1998. This book is slightly different because the target audience for the LEGO Spybotics product is much younger than the one I usually write for, which is tricky because I find that I need to strike a balance between keeping older readers interested without swamping younger readers with information. Kids are very aware of their age, so windows of approval for books, movies, and games are narrow indeed. That's why kids suddenly dislike things that they've loved for years.

I encouraged quite a few children to read this book as it went through different drafts. Of course, the chance to play with a new LEGO product made things much easier. My twin 10 year olds, Eric and Owen, constructed the Spybots on their own from the instructions and I noted any difficulties they had. In almost all cases, the problems were due to hasty assembly and not looking carefully at the instructions on the screen. Even Graham, my 6 year old, was able to build a Spybot with just a bit of help.

LEGO was very generous in providing one of each of the Spybots for evaluation during the writing phase. Their technical team was helpful when I had questions about the product. Fortunately, the typical attention to detail of the LEGO Spybotics development team ensured that there were very few of these questions.

The editorial and production team at Apress was a pleasure to work with, and gave me the opportunity to make this book a reality under incredibly tight production deadlines. Thanks go to Gary Cornell for taking a chance on a book that is a little outside the boundaries of the typical Apress product. Grace Wong and Nicole LeClerc provided welcome editorial corrections that improved the focus and flow of the information. Kurt Krames stepped up to the plate and hit one out of the park on the design of this book. The marketing and public relations team at Apress, Stephanie Rodriguez and Hollie Fischer, have done a wonderful job selling and promoting this book on short notice. David Koudys did the technical review of the book to make sure there were no glaring inaccuracies. Any remaining errors are entirely my own.

Once again, I have to thank my wife, Christine, for her support during yet another crazy summer of writing, testing, and revising, and the inevitable writer's block. Yes, technical authors also suffer from this affliction. Fortunately, it is often cured by an afternoon away from the computer and in the cool, clear, and clean sand and water of nearby Sauble Beach, Ontario.

PREFACE

The LEGO Spybots are an exciting addition to the world of LEGO elements. They cannot be put into a single category of toys because they draw from the best parts from many different types of play. They have an appeal that is difficult to describe but easy to understand from the moment you watch the introductory video on the CD-ROM included with each kit. While the product is aimed at kids ages 9 to 14, younger and older folks can have a lot of fun with Spybots too.

Who This Book Is For

You do not need any special skills in mechanics, electronics, or programming to enjoy LEGO Spybots. The premise is that an organization called Secret Missions Agent Robot Teams (S.M.A.R.T.) uses teams of Spybots and humans to solve problems in hotspots around the world. As an agent, you can choose from one of ten predesigned missions and earn points toward a higher security rating and bragging rights among other agents. You operate the robot using a handheld controller, and missions range from breaking secret codes to dodging a laser maze in an old factory. You connect your Spybot to the host computer to update your rankings. If you have an Internet connection, you can rank yourself against other agents around the world.

This book is designed to help you get the most out of the LEGO Spybotics product. It is written for a reader who is about 12 years old, but even 9-year-old readers should have no problem with most of the text. Younger readers may need help with some of the technical discussions, but I have tried to include physical experiments that readers can do to reinforce the concepts. Either way, you can enjoy the Spybots right out of the box, and this book will give you additional information to help you understand the product better in the following ways:

- Provides detailed descriptions of Spybot hardware systems
- Gives you basic background information on the Spybot electronics
- Includes tips to help you solve common problems
- Supplies detailed mission summaries
- Shows you how to program your Spybot to run missions by itself

A lot of the information in this book is available on the LEGO Spybotics CD-ROM, but it can take a lot of time to find it. This book will put all of the critical information you need as a S.M.A.R.T. agent right at your fingertips. You can even read it when you're not running missions with your Spybot!

Spybotics Training Manual Overview

This book goes into more detail than is available in the online documentation. This is one of the first LEGO products that has most of the documentation, including building instructions, on a CD-ROM. You can get a lot of the information you need from the video and audio tracks on the CD-ROM, but this book connects the points that are often in separate places on the CD-ROM. The LEGO company has done a great job of putting together these animations—this book is meant to serve as an additional offline reference.

This book has seven chapters, each covering a specific topic, and one appendix. You can use the LEGO Spybots right out of the box, and this is probably what most readers will do at the beginning. You can probably find most of the basic information you need by browsing around on the CD-ROM, and reading this book from start to finish will give you additional, in-depth information.

Chapter 1: Spybotics Overview is a general introduction to the LEGO Spybotics product line and provides a summary of the minimum computer hardware requirements and an installation guide.

Chapter 2: Spybot Mechanical Systems Training is your field guide to the common mechanical features of the Spybotics system. It also provides specific information on the unique aspects of each Spybot.

Chapter 3: Spybot Electronic Systems Training is where you'll find detailed information on the electronic systems your Spybot has on-board. Don't worry if you aren't technically inclined—the explanations are written in clear language.

Chapter 4: Spybot Construction and Care is aimed at readers who have not built anything with LEGO Technic elements before. However, even experienced builders will find some hints for Spybot construction in this chapter.

Chapter 5: Spybotics Agent Communications shows you how to communicate with your Spybot and extract mission results for comparison with other agents. There is also a section on using the Spybotics Web site to exchange secret information and missions with other agents.

Chapter 6: Mission Selection Guide gives an overview of how to set up your mission space and how to load the mission into your Spybot(s). Also included is a brief description of each mission and settings that you can use to change the mission difficulty.

Chapter 7: Mission Customization Guide shows you how to change the way your Spybot works. After all, this is a LEGO product. You'll also learn some simple programming that you can do to create solo behavior for your Spybot.

Appendix: Mission-Specific Capsule Cross-Reference describes the capsules that are available only in certain missions. You can use these capsules to give your Spybot special powers during missions. The capsules are cross-referenced to each mission and profile.

Additional Spybotics Resources

Interested agents should visit http://www.hempeldesigngroup.com/lego/spybotics for Spybotics news items and links to stories of interest to Spybotics agents. Also, don't forget to visit the official LEGO Spybotics site at http://www.spybotics.com for additional tools and games, and to read about the history of Spybotics.

Note to Parents

The LEGO Spybots encourage play on many different levels. You start with the traditional LEGO construction phase, where you learn about the simple mechanisms that make the Spybots run, and then you progress to driving the Spybot around using the remote control. This by itself leads to all kinds of free play. The next step is to investigate the mission maps on the CD-ROM, where there are all kinds of opportunities to investigate the world around you, from the bounty of the South American rain forest to the effects of global warming on the polar ice caps.

The Spybot missions encourage cooperation and thoughtful execution, rather than the more common computer game themes of combat or mindless hack-and-slash rampages. Even the missions that involve two or more Spybots involve themes such as rescue of a disabled comrade and the prevention of data theft. As a parent myself, I appreciate the work that went into the design of the missions and the lessons they can help reinforce.

The imagination is exercised in a subtle way when playing with LEGO Spybots. The mission briefs on the CD-ROM encourage the agent to pretend that a lamp is an energy resource or that an obstacle is an electrical impulse generator. Some kids will naturally want to build and decorate their own obstacles and mission area, which can lead to all kinds of useful skills development.

Most important, it's a toy that parents can enjoy with their children. It's pretty hard to resist playing with a remote control vehicle that has preprogrammed missions with specific goals—especially when your mission fails and you hear yourself say that you will try "just one more time."

1

SPYBOTICS OVERVIEW

The Spybotics system from LEGO is a completely new and revolutionary type of toy. It combines the well-known LEGO interconnecting elements, a preprogrammed robot module, and the features of a video game. The result is nothing short of spectacular because it moves the play area of the game out into your living space.

This chapter covers the following topics:

- Introducing Spybots
- Understanding the computer system requirements for the Spybotics software
- Installing the Spybotics software
- Installing the Spybot communication cable

Spybot Introduction

What exactly is a LEGO Spybot and what do you need to know to put one together? It might be easier to start with what a Spybot isn't and what you don't need to know—especially if you're intimidated by the complexity of the LEGO MINDSTORMS intelligent bricks.

- Spybots are not general-purpose robotic building elements.
- Spybots do not require any programming knowledge.
- Spybots do not require advanced knowledge of mechanical design.
- Spybots do not require any advanced electronics knowledge.

At the core of each Spybot is the same intelligent *brick*, with a visible light sensor, a touch sensor, infrared communications and range-finding capabilities, and two built-in motors. You'll learn more about these systems in Chapters 2 and 3. There are currently four different Spybots, as shown in Figure 1-1, each with its own color and external mechanical design. The Spybot kits come with the main brick, a handheld remote controller, all of the parts for the Spybot, and a CD-ROM with on-screen building instructions.

▲ **Figure 1-1.**
The Spybot family: Snaptrax, Technojaw, Shadowstrike, and Gigamesh

The CD-ROM also contains the software that turns the Spybot into a fun game that the whole family can enjoy. You're teamed up with one or more Spybots and your goal is to complete missions for points that go toward increasing your Secret Missions Agent Robot Teams (S.M.A.R.T.) security clearance. The more you practice, the better you get, and you'll learn that sometimes it's best to proceed carefully. Going into a secure installation with guns blazing isn't always a good way to get the job done.

There are ten predesigned missions that you and your Spybot can complete to earn points toward increasing your security level. You can set the complexity of each mission to make it easier or harder to accomplish. The tougher the mission, the more points you earn toward your security rating if you succeed. As you increase your security level, you see your individual ranking increase. It's easy to check what your security clearance is—just turn on your Spybot, see how many lights come on, and count the number of beeps!

But it gets even better. If you choose to connect with other Spybotics agents on the LEGO Spybotics Internet site (http://www.spybotics.com), you can compare your progress against theirs.

You can do all of these things without ever having to know about programming, mechanical systems, or electronic systems.

This book will help you learn about how to use your Spybot effectively. You'll find out about the Spybot mechanical design features, the Spybot electronic systems, and how to use them. You'll even get some hints for building your Spybot if you've never played with LEGO toys before. I've also included mission summaries (and tips for improving your score) so that you can be fully prepared to execute missions with your Spybot partner. Finally, this book includes a chapter (Chapter 7) that will help you when you try your hand at designing your own missions.

Spybotics System Requirements

The computer requirements for the LEGO Spybotics system are fairly modest. The Spybotics system software only runs on Windows machines. You will need

- 350 MHz or faster processor

- 64MB RAM

- 200MB of free hard disk space (slightly more space is required if QuickTime, Flash, and DirectX installation are needed, and a full installation that lets you play without the CD-ROM in the computer will use 600MB)

- 16-bit color (or more) Super VGA monitor

- 4MB graphics card (DirectX 8.0 compatible)

- Sound card (DirectX 8.0 compatible)

- One free 9-pin RS-232 serial port

- Optional: 28.8 Kbps modem and an Internet connection to get to S.M.A.R.T. Headquarters online

- Windows 98/98 SE/ME or Windows XP

Each Spybot needs three AA (LR-06) batteries, and the controller uses three AAA (LR-03) batteries. You can find out more about the different kinds of batteries in Chapter 4.

According to the official LEGO Spybotics documentation, Windows 2000 is not supported, but many agents, including myself, have installed the Spybotics software on Windows 2000 machines. If you're not too concerned with viewing all of the neat animations, you might be able to get away with a slower computer as your connection with S.M.A.R.T.

I've installed the system on an old Pentium 233-MMX laptop and have acceptable performance.

Installing the Spybotics Software

To install the Spybotics software on your PC, just insert the CD-ROM into the drive on your computer. The installation procedure starts automatically. If it doesn't (you might have AutoPlay disabled), then you can run the "setup.exe" program directly from the CD (click Start, select Run, and then type **x:\setup.exe**, where **x:** is the drive letter for your CD-ROM).

During the installation, you are given the option of a partial or full install. The partial install takes about 200MB of disk space and the full install takes about 600MB. If you have lots of disk space, I suggest a full install because it allows you to play without having the Spybotics CD in your CD-ROM drive.

> **TIP** Eventually, you may find yourself short of hard disk space. If you remove the Spybotics application using the uninstall shortcut, the mission settings and agent rankings are saved on the hard disk. When you reinstall the application, your old settings and rankings are also restored.

When the application is installed, you are greeted by a short introductory video and then a picture of all four Spybots on the S.M.A.R.T. Headquarters screen (see Figure 1-2).

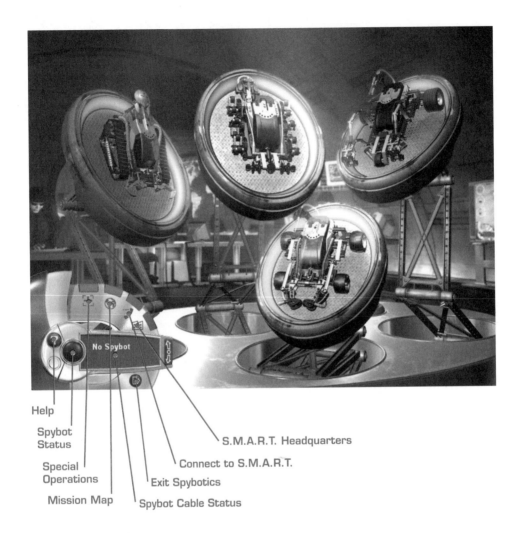

Help

Spybot
Status

Special
Operations

Mission Map

S.M.A.R.T. Headquarters

Connect to S.M.A.R.T.

Exit Spybotics

Spybot Cable Status

▲ **Figure 1-2.**
The S.M.A.R.T. Headquarters main screen

Click the image of your Spybot to bring up its construction files. If you're eager to start building your Spybot partner and you're new to LEGO or LEGO Technic construction, I suggest you skip to Chapter 4 to review some basic building tips. I'll wait for you to finish building before moving on.

Connecting Your Spybot to Your Computer

The next step is to connect the Spybot communications cable to a spare 9-pin serial port on the back of your PC (see Figure 1-3). You might need to get some help from a senior agent (like a parent) to help you with this. Power down the computer and connect the 9-pin end of the cable to the matching port on your PC.

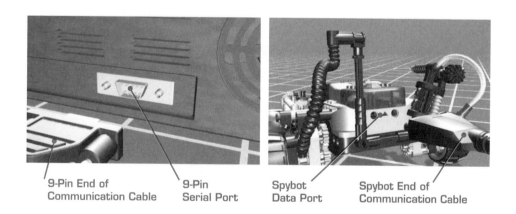

9-Pin End of
Communication Cable

9-Pin
Serial Port

Spybot
Data Port

Spybot End of
Communication Cable

▲ **Figure 1-3.**

Connecting the communications cable to the computer and the Spybot

If you don't have a spare serial port, you have a couple of options.

- If you're using a serial mouse, consider getting a PS2 or USB mouse.

- If you're using an external modem that uses the serial port, consider getting an internal modem.

- If you have only a 25-pin serial port, consider getting a 25- to 9-pin adapter.

- If you don't actually have a serial port, consider getting a USB serial adapter.

Refer to the printed leaflet that comes in the Spybot kit for details, or run the Trouble-shooter on the CD-ROM if you have questions at this point. You can get to the Trouble-shooter by clicking the gray Spybot Status button on the communicator in the lower-left corner of the screen, and then clicking the button at the upper-right of the Spybot Status screen labeled "Trouble-shooter" (see Figure 1-4). Note that you can call up the Status screen at any time and it will overlay the current screen. You cancel this screen by clicking the "X" button in the upper-right corner.

If you are still having problems with the Spybotics application, you can visit the main Web site at http://www.spybotics.com or read the support document that you can find on your computer here: C:\Program Files\LEGO Software\Products\Spybotics\EnglishSupport\Readme.txt.

You can tell the cable is correctly installed when the communicator in the lower-left part of the screen shows the message "No Spybot." If the cable is not connected or cannot be found, the message will read "No Cable."

If you jumped ahead and built the Spybot before you got the communication cable properly set up, you'll have a chance to do it when you get to the end of the building instructions. There is a complete systems check at the end of the construction phase.

Status: Gives you overall system status, such as battery life, clearance attained, and overall time and experience of specified Spybot.

Technical Manual: Shows you specific functions of Spybot remote and unit, as well as test programs to diagnose your Spybot and controller.

Trouble-shooter: Diagnoses where a problem may lie if the Spybot doesn't function properly, and verifies connectivity from PC to Spybot and IR signal communication from Spybot to remote.

Agent Points: Gives you specific results from missions.

▲ Figure 1-4.
The Spybot Status screen overlay

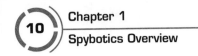

Summary

At this point you should have the LEGO Spybotics software installed, and you may have skipped ahead and built and tested the Spybot. You're now ready to move on and learn more about your LEGO Spybot partner. Next you'll find out about how the mechanical parts work together with the electronic systems in an easy-to-understand format.

The more you know about your Spybot, the better you will be as a team.

SPYBOT MECHANICAL SYSTEMS TRAINING

Your choice of Spybot can make the difference between mission success and failure. This chapter covers some of the features common to all Spybots and the features that make each Spybot unique. You'll also find some ideas to consider when you develop your own Spybots.

This chapter covers the following topics:

- Identifying the common features of each Spybot
- Determining the relative speed and strength of a Spybot
- Describing the special features of each Spybot

One of the techniques I use to explain the Spybot features and operation is simple exercises designed to help you understand what's going on by thinking and acting like a Spybot. Experienced designers and engineers will often do this to get a better grasp of the problems they might have with their projects.

Common Spybot Features

The basic LEGO Spybot brick is the same (except for the color of the transparent domes) on all of the Spybots. Chapter 3 goes into more detail about the electronic systems in your Spybot. In this section, I just talk about the mechanical aspects of the Spybot brick.

Each Spybot brick has a battery compartment, a touch sensor, and two motors, as shown in Figure 2-1. Each of these mechanical features either provides input to or is controlled by the computer inside the Spybot brick.

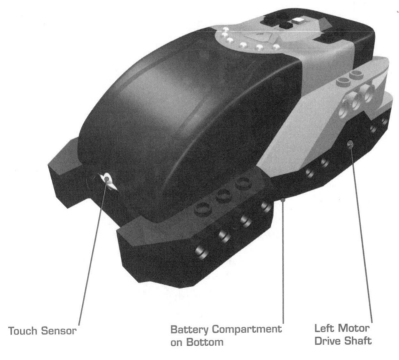

Touch Sensor   Battery Compartment   Left Motor
                        on Bottom   Drive Shaft

▲ **Figure 2-1.**
Spybot brick mechanical features

If you look closely at your Spybot and/or the Spybot in Figure 2-1, you'll notice the following additional features:

- **Two buttons on top:** The black button is the power button, and the silver/gray button is the Run button to run the current downloaded program.

- **Six lights in an arc:** These lights are for display. During missions, various lights display for the S.M.A.R.T. agent to make appropriate responses. The lights also indicate the status of a mission. When you first turn the Spybot on, the number of lights displayed indicates your current Spybotics security clearance.

- **Center yellow light:** This light displays information during missions and shows you that the Spybot is ready to start a mission.

You will learn more about these additional features in the appropriate sections of Chapter 3.

The Battery Compartment

On the underside of the Spybot brick, you'll see a panel with beveled edges and an arrow molded into the part. Gently push the panel in the direction of the arrow and lift it clear of the battery compartment. There is room for three AA (LR-06) batteries. For more information on batteries, see the "Choosing Among Battery Alternatives" section in Chapter 4.

When you insert batteries into the compartment, be careful to ensure that the "button" on the top of each battery is pointing in the right direction. There is a picture molded into the bottom of the compartment that will help you get it right. Remember, you can't hurt the Spybot by inserting the batteries the wrong way, but it won't work if you do so.

When you replace the panel, make sure the arrow is pointing to the front of the Spybot (see Figure 2-2). You should also make sure that the front of the panel is in line with the front of the Spybot. Then gently push the panel back to seal the battery compartment.

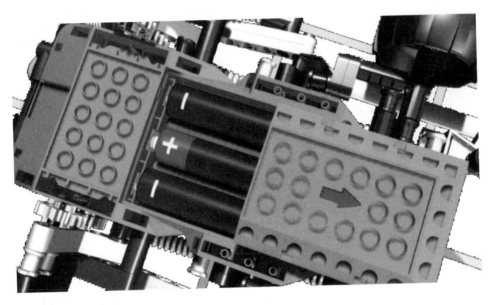

▲ **Figure 2-2.**
Spybot brick battery placement

If you're thinking about building a custom Spybot, it's good to keep the access to the battery compartment free of any obstructions. Avoid running bracing or suspension parts across the battery compartment.

The Touch Sensor

The Spybot *touch sensor* is used to detect when the Spybot has hit something. It looks simple by itself, but it can be quite a challenge to build a good bumper for the sensor. Although the sensor is designed with a "shock absorber" to minimize damage when it is fully pressed, it's not a good idea to allow your Spybot to simply crash into things.

For one thing, the touch sensor is very narrow. This means you have to be pretty lucky to hit anything at all. Another thing to watch out for is that the Spybot models generally have wheels or tracks that extend past the front of the brick. This makes it even harder to activate the touch sensor.

All Spybots have a bumper, and as you might expect, S.M.A.R.T. has gone to a lot of trouble to design bumpers that are easy to activate and yet protect the Spybot brick. Here's an exercise you can do to prove to yourself that a flexible yet strong bumper is a good thing to have.

EXERCISE

Spybot Bumper Design

Imagine that your nose is a touch sensor. What will happen if your target is the wall and you have no bumper to protect your nose? You probably don't need to do this part of the experiment to come up with the answer, right?

Now put your arms straight out in front of you and lock your elbows (see Figure 2-3A). If you walk toward the wall, your hands will hit the wall and the shock will be transmitted right back to your shoulders. This time your nose is protected, but your arms are not flexible enough to prevent possible damage to the rest of your body.

A   B

▲ **Figure 2-3.**
Experimenting with bumper designs

Now put your arms in front of you and bend your elbows slightly (see Figure 2-3B). When you get to the wall, your arms flex, which takes up almost all of the impact. As your arms flex, your nose gets closer to the wall and will eventually touch the wall gently.

As you can see, even something as simple as a bumper needs careful design. The "Individual Spybot Features" section later in this chapter describes each bumper in more detail.

The Drive Motors

There are two motors on each Spybot. You can direct each motor individually from the handheld controller. On a normal remote control car, you have one motor that drives the vehicle forward and backward, and another that steers the vehicle. Why don't the Spybots use this arrangement?

The answer is simplicity and cost. The S.M.A.R.T. design team realized that a conventional steering system would have more parts and would be more likely to break. Like all good designers, they looked around for a system that was simple to build and operate. The system they decided to use is called a *differential drive* mechanism—one motor moves one side of the vehicle, and the other motor moves the other side. You'll most often see this kind of mechanism on construction equipment, where strength and reliability are important.

You can find the differential drive system in all kinds of equipment, including bulldozers (see Figure 2-4), skid-steer loaders, huge excavators, and even commercial lawn mowers. This type of system has the advantages of being very simple to build and easy to control. But there is one disadvantage—can you think of what it is?

▲ **Figure 2-4.**
LEGO Set 914 bulldozer (courtesy of Coby at http://tubafrog.com/lego)

If you guessed *friction,* you're right. If you guessed another exercise is coming up, you're right again.

EXERCISE

Friction

To understand how differential drive works and how it's affected by friction, stand on a smooth floor with just socks on your feet. When you move, take small steps by taking some weight off one foot and shifting that foot forward or backward slightly before putting it down again. Try to keep your feet about shoulder-width apart and parallel.

Moving forward and backward is no problem. To turn to the left, keep your left foot on the ground and move your right foot forward a few times. See how your left foot spins on the smooth floor? This is where the friction comes from. You've got socks on, and the floor is smooth, so it's easy for your foot to skid in the turn. If you've ever run down a long hall in socks and skidded, you know what I'm talking about.

Now put clean sneakers on your feet and try to turn. Remember to keep your shoes flat on the floor. Don't lift the heels or toes of the shoes. The rubber soles on the sneakers have better contact with the smooth floor, so they have more friction. It's actually hard to turn one of your feet unless you reduce the contact patch by raising the heel or toe. That's why basketball players lift their heels when they pivot to look for some-one to pass the ball to. Friction is also the reason you can't slide down that long hallway with new sneakers on your feet.

The main idea here is that a machine with differential drive steers by turning only the drive wheels or tracks on one side. You don't actually point the wheels in the direction of the turn like you do with a bicycle or a car. Now that you have an idea of how differential drive or skid steer-ing works, you can probably find other examples of machines that use this control method.

Determining Spybot Speed and Strength

Each LEGO Spybot has a different arrangement of gears, wheels, and tracks that allow it to move freely. All of the Spybots except for Snaptrax use gears to transfer motion from the motor to the drive wheels.

But how do the gears affect the speed and strength of the Spybot? This section of the book will help you understand how gears work, and it will also provide design information that Spybot builders can use for their own creations.

The Spybots use the simplest possible gear train. The motor shaft drives the input gear, which drives the output gear, which is connected to the wheels (see Figure 2-5). The Shadowstrike, Technojaw, and Gigamesh units each have a small gear driving a larger one. The Snaptrax drives its output wheel right from the motor.

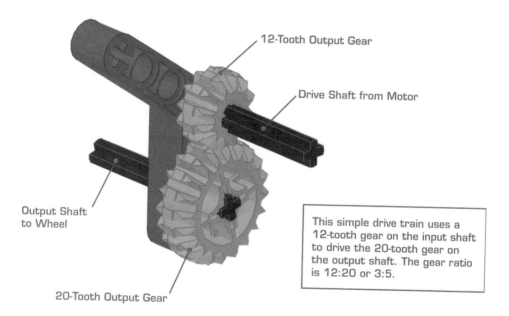

12-Tooth Output Gear

Drive Shaft from Motor

Output Shaft to Wheel

This simple drive train uses a 12-tooth gear on the input shaft to drive the 20-tooth gear on the output shaft. The gear ratio is 12:20 or 3:5.

20-Tooth Output Gear

▲ Figure 2-5.
Simple Spybot gear train

Before you can figure out how fast the Spybots will go, you need to do a bit of background work using simple math. Don't worry if you think you're not good at math. By the time you finish this section, you'll be an expert on measuring circles and figuring out gear ratios.

Measuring Circles

The time has come to do another exercise. This time, you're going to figure out the relationship between the distance across a circle and the distance around a circle. If you already know the answer, you can skip this exercise.

EXERCISE

Distance Across a Circle vs. Distance Around a Circle

For this exercise, you'll need the following items:

- A few round household items (soda can, coffee mug, CD, and so on)

- A piece of string large enough to go around the biggest item

- A ruler

- A calculator

First, write down the names of the household items you selected in the Household Item column in Table 2-1. Wrap the string around each of the household items and keep track of where the string overlaps. This distance is the *circumference* (the distance around) the object.

Measure the distance from the end of the string to this overlap point and write down the result in the Circumference column in Table 2-1.

Then measure across the widest part of the item to get the *diameter* of the item and write that number down in the Diameter column in Table 2-1. Don't worry about the Ratio column just yet. I've filled the first row in to get you started.

▼ **Table 2-1.**
Circle Measurement Results

HOUSEHOLD ITEM	CIRCUMFERENCE	DIAMETER	RATIO
Dessert plate	20.875 in. (530 mm)	6.625 in. (168 mm)	3.154

It's a little easier if you use millimeters (mm) for the measurement. If you use inches (in.), you might need to get a grownup to help you convert the fractions into decimal numbers. For example, if you measure 8 3/4 inches for the circumference of an item, write down **8.75 in.** in the Circumference column.

Now, to fill in the Ratio column, divide the circumference by the diameter using your calculator. Write down the result for each row in the last column. The interesting result is that the numbers should all be pretty much the same, depending on how carefully you measured the circumference and diameter.

What you have just done is figured out an approximate (rough) answer for the value of π (pi). Without getting into a lot more detail than you need right now, the real value of π is about 3.14159.

No matter what the size of the circle, you can always figure out the distance around if you know the diameter by multiplying it by π. On the other hand, if you know the distance around the circle and you want to know the diameter, you just divide by π. Figure 2-6 shows you how to calculate the circumference and diameter of a circle with π.

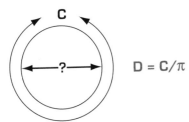

$$D = C/\pi$$

C is the circumference
D is the diameter
π is about 3.14159

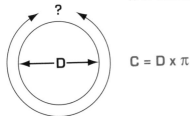

$$C = D \times \pi$$

▲ **Figure 2-6.**
Calculating the circumference and diameter of a circle

So what does all this have to do with how fast a Spybot can travel? Read on to find out.

Speed and Wheel Size

Now that you have a simple way of figuring out how big around a circle is (its circumference), you can move on to calculating speed. Have a look at the small wheel (the wheel at the top of the diagram) shown in Figure 2-7. If the small wheel has a diameter of 10 inches, how far will it travel in one revolution? (Hint: The distance around the wheel [circumference] is how far the wheel will travel in one revolution.)

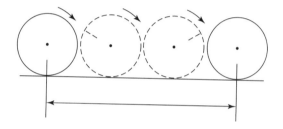

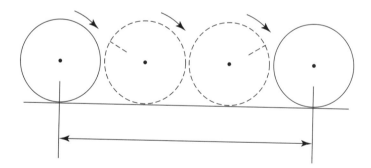

▲ **Figure 2-7.**
Which wheel moves farther in one turn?

If you answered 31.4159 inches, you're right! Now look at the big wheel (the wheel at the bottom of the diagram) in Figure 2-7. It has a diameter of 15 inches, so it will travel 47.1239 inches in one revolution. OK, this is the cool part. Let's say each wheel turns at the same speed of one revolution per second. An engineer would say that each of the wheels has an *angular velocity* (turning speed) of one rotation per second. If you put these wheels side by side, how far does each one travel in 1 second?

The small wheel goes 31.4159 inches in 1 second, and the big wheel goes 47.1239 inches in 1 second. So not only does the big wheel go further in one turn, it also goes forward faster!

Hang on, you're almost ready to put this all together to calculate Spybot speed. You just need one more bit of information, and that's how the gears change the speed of rotation of the motors to the wheels.

Spybot Gear Trains

All of the Spybots use a simple arrangement called a *gear train* to translate the rotation of each motor's axle to the drive wheel. The Spybot sets use the newer LEGO gears that can be used inline or at right angles to each other. If you have any of the LEGO BIONICLE sets, you're probably familiar with this already.

The gear trains used in Spybots are very easy to understand. If you have a bicycle that has adjustable gears, you already have a good idea of how they work. Let's say you have a bike with a six-speed block on the back wheel and a single fixed gear on the chain ring (see Figure 2-8). To go faster, you make the chain go on the smaller gears, but then it's harder to turn the pedals. To go up a hill, you put the chain on the bigger gears, but then you travel more slowly.

Output Gear (6 Speeds)   Input Gear

▲ **Figure 2-8.**
Parts of a bicycle gear train

> **★ NOTE** Engineering is the art of trading one property for another to achieve a goal. When you ride your bicycle and change gears to go faster or to climb hills, you are making decisions about trading speed for power. The more you understand how these things work, the better your ability to make these kinds of engineering decisions.

It's time for another exercise. This time you'll use the Technojaw gear train as your example.

EXERCISE

Calculating a Simple Gear Ratio

As you can see in Figure 2-9, Technojaw's 12-tooth gear is attached to the motor drive shaft, and its 20-tooth gear is attached to the wheel axle.

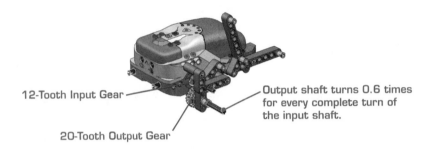

12-Tooth Input Gear

Output shaft turns 0.6 times for every complete turn of the input shaft.

20-Tooth Output Gear

▲ **Figure 2-9.**
The Technojaw gear train

A bit of thought will help you figure out how far the axle turns for every motor revolution. Let's say that the gear on the motor axle turns by one gear tooth. How many teeth does the gear on the wheel axle turn by? Also one tooth. So after the motor gear turns by one full revolution (12 teeth), the wheel gear has also rotated by 12 teeth. But the wheel gear has 20 teeth, so it has only turned by 12 out of 20 teeth. That's 12/20 or 3/5 or 0.60 revolutions for every revolution of the motor.

Calculating Spybot Speed

Now you can combine your work on circles, wheel speed, and gear trains to figure out the relative speeds of the Spybots. Because I'm an engineer, I like to make tables for this kind of calculation because then I can see all of my work. Creating a table also helps because the calculations are basically the same for all the Spybots.

You're going to figure out the relative speeds of the Spybots, so you start by assuming that the motors are all the same and turn at one revolution per second. Then you use the gear-train ratio to figure out the drive axle speed. Finally, you multiply by the wheel circumference to get the relative speed of the Spybot. I've worked out the answer for Shadowstrike in Table 2-2 to get you started.

▼ **Table 2-2.**
  Spybot Speed Calculation Results

SPYBOT	MOTOR GEAR	AXLE GEAR	RATIO	WHEEL CIRCUM-FERENCE	SPEED
Technojaw	12	20	_____	_____	_____
Snaptrax	direct drive	direct drive	1	3.125 in. (80 mm)	3.125 in. (80 mm) per motor rev
Gigamesh	12	36	_____	5.44 in. (138 mm)	_____
Shadowstrike	12	20	0.60	6.25 in. (158 mm)	3.75 in. (94.8 mm) per motor rev

The wheel circumference is a bit tricky to calculate for some of the Spybots. The easiest way to get the wheel circumference is to make a little mark on one of the tires and place the mark on the tire at the end of a line drawn on a sheet of paper. Roll the tire until the mark touches the paper again and measure the distance between the marks. That's the circumference. The Snaptrax wheel circumference is difficult to calculate because it's a track, so I've filled that in for you in Table 2-2 also.

If you assume that all of the motors spin at the same speed, you can already see that Technojaw is slightly faster than Snaptrax. Can you figure out the speed of the rest of the Spybots? After you've filled out the table, check to see if your answers are correct by looking at the http://www.hempeldesigngroup.com/lego/spybotics Web site.

As you'll see in the next section, speed isn't everything. Each Spybot has its own special features that make it suited to particular tasks.

Individual Spybot Features

The Spybots are all capable of detecting light, moving in all directions, and figuring out when they have hit an obstacle with their front bumper. But sometimes you'll need special features that give your Spybot the edge. Let's look at each Spybot and figure out what makes it unique.

I start by describing the simpler designs, Snaptrax and Gigamesh, and then I move on to Technojaw and Shadowstrike.

Snaptrax S45 Features

The Snaptrax Spybot is simplest of all of the Spybots (see Figure 2-10). It has no gearing between the motors and the drive track, just a special drive axle that has a shoulder to keep it in place. The axle is unique because it is made of a softer plastic than the normal LEGO axles. This helps to protect the motor from shocks.

▲ **Figure 2-10.**
The Snaptrax S45 Spybot

The Snaptrax is relatively narrow and is well protected against attack from other forces. The tracks have guardrails to keep obstacles away, and the snapping claw arrangement on the front is combined with a simple but effective bumper to protect the main brick.

Just about the only disadvantage to Snaptrax is that you have to be careful about how you control it. If you drive backward and then quickly change to driving forward, you can make Snaptrax "pop a wheelie," which makes it more difficult to control and also more vulnerable to attack.

Gigamesh G60 Features

The Gigamesh Spybot is certainly a powerful-looking device (see Figure 2-11). With its strange, three-pronged wheels, it can find traction on even the bumpiest terrain. Because of its extremely low gear ratio, it is also the slowest Spybot, but it makes up for its slow speed in brute strength.

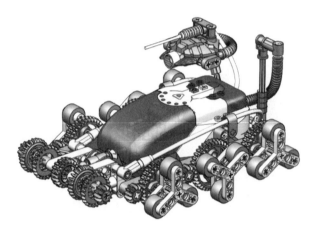

▲ **Figure 2-11.**
The Gigamesh G60 Spybot

The Gigamesh is also a narrow Spybot, and it is very stable because it is low to the ground. It has a simple bumper arrangement, and it has an intimidating set of grinding gears along its front edge. With all of the gears that come with Gigamesh, you can be assured that other building ideas will come to you.

One of the disadvantages of Gigamesh is the complexity of its gears. One or two times, I've had loose LEGO elements get caught between Gigamesh's Y-shaped wheel holders. The only way I found to clear the blockage was to reverse direction and twist away from the obstacle(s).

Technojaw T55 Features

Technojaw is a very interesting Spybot because it combines speed with an unusual "pinching" action of its front claws (see Figure 2-12). If you look closely at the drive mechanism, you'll see that if you drive Technojaw forward, the drive train pulls the jaws open. If you stop Technojaw, the jaws close automatically.

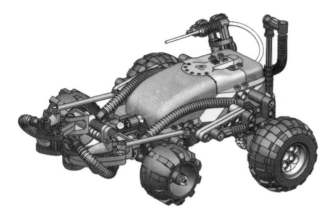

▲ **Figure 2-12.**
The Technojaw T55 Spybot

Technojaw is quite wide. This is to make it easier to accommodate the pincher mechanism within the protection of the balloon tires. The plastic tires at the front seem a bit out of place, but an experienced Spybot agent will soon figure out why the designers didn't use rubber. The reason is because the tires act as skids when Technojaw turns. The rubber wheels would have too much friction and it make it very difficult to turn Technojaw.

Technojaw has a great assortment of parts that are useful in other Technic creations. If Technojaw has one disadvantage, it's width. The width of this Spybot makes it difficult to maneuver in tight situations.

Shadowstrike S70 Features

Shadowstrike is a very cool Spybot (see Figure 2-13). It has the same type of pinching action as Technojaw, but it's much more exaggerated. Shadowstrike also has an unusual three-wheeled configuration that combines a skidding wheel with a bumper. The very narrow bumper and deflection beams make Shadowstrike less likely to hit obstacles and register them on the touch sensor.

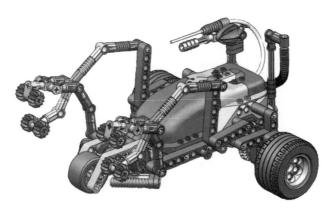

▲ **Figure 2-13.**
The Shadowstrike S70 Spybot

Shadowstrike is also a wide robot, but its narrow front profile makes up for it. Shadowstrike has a good assortment of beams and axle connector parts, which makes for interesting secondary models.

If there is one disadvantage to Shadowstrike, it's the complexity of the gripper mechanism. In close combat with another Spybot, there is always the danger that Shadowstrike's gripper will get caught on something.

Summary

In this chapter, I went over the basic physical and mechanical features in each LEGO Spybot and gave you some detailed engineering calculations that you can use to compare Spybot speeds. Along the way, you've learned a few things about gears and drive trains too. I compared the individual Spybots so that you can make the important decision of which should be your first Spybot. If you already have one, then you have more information for deciding which Spybot will be the *next* member of your Spybot team!

SPYBOT ELECTRONIC SYSTEMS TRAINING

In this chapter, you'll learn about the electronic systems that are common to all of the LEGO Spybots. Pay close attention, Agent. This chapter has information that will allow you to join up with other agents to complete missions.

This chapter covers the following topics:

- Using the Spybot Status screen and Technical Manual

- Examining the Spybot pushbuttons, indicator lights, and handheld controller/beacon modes

- Using the Spybot's infrared range-finding capabilities

- Examining the data port, light sensor, and laser port

- Understanding the default behavior of the Spybot

By the end of this chapter, you'll have a very good understanding of the basic operation of the Spybot electronic systems, and you'll be ready to build and use your Spybot on missions.

Using the Spybot Status Screen and Technical Manual

The LEGO Spybotics CD-ROM has a wealth of information on it. One of the most important areas is the Spybot Status screen. From the S.M.A.R.T. home screen, just click the gray Spybot Status button to bring up the Status screen (see Figure 3-1).

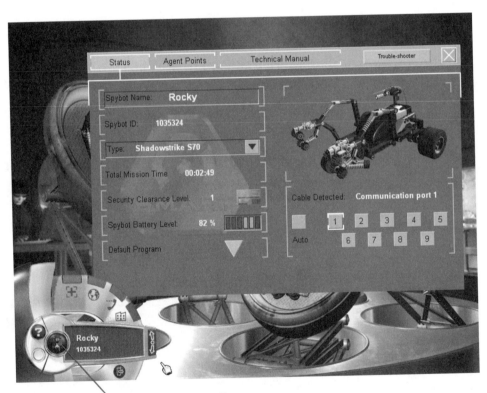

Spybot Status Screen Button

▲ **Figure 3-1.**
The Spybot Status screen

On this screen, you are able to do a number of things, including

- Give your Spybot a name (useful if you have more than one Spybot!)
- Check your Spybot's unique factory set ID
- Set the Spybot model type
- Check how much mission time your Spybot has in the field
- Check the security clearance level you have achieved with your Spybot
- Check the Spybot battery level
- Download a default program to your Spybot partner
- Check and set the serial port the Spybot cable is connected to

You should get used to examining this Status screen every time you start the game up or before you program your Spybot. On this screen, you can easily keep track of your Spybot's progress and check out the battery level in your Spybot. You can find out more about the Spybot Status screen in Chapter 5.

The Spybotics CD-ROM has a lot of the information in this chapter in video form in the Technical Manual. To access the Technical Manual from the Spybot Status screen, click the Technical Manual button (see Figure 3-2). You don't need to have the Spybot built in order to view the Technical Manual.

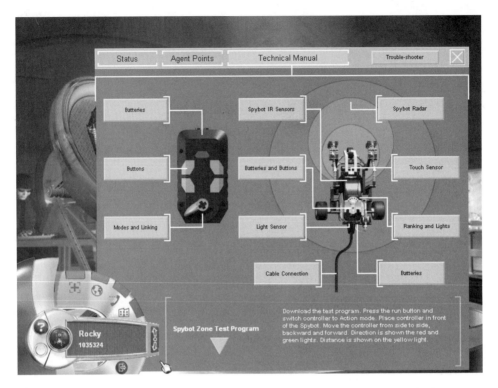

▲ **Figure 3-2.**
The Spybot Technical Manual screen

======================== EXERCISE ========================

Downloading the Default Zone Test Program

Before you read the rest of this chapter, download the Default Zone Test
Program by following these steps:

1. Start the Spybotics software.

2. Turn on your Spybot and connect it to the serial cable.

3. Click the Spybot Status button.

4. Click the Technical Manual button at the top of the Status screen.

5. Click the Spybot Zone Test Program button at the bottom of the Technical Menu screen.

This default program lets you drive your Spybot using the handheld controller and investigate the infrared range-finding capabilities.

Identifying the Spybot Electronic Systems

Before I go into too much detail on the Spybot electronic systems, you need to know the names of the parts in the system (see Figure 3-3).

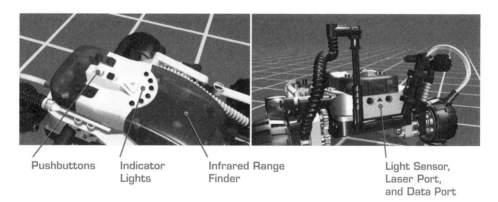

Pushbuttons   Indicator Lights   Infrared Range Finder   Light Sensor, Laser Port, and Data Port

▲ Figure 3-3.
The Spybot electronic systems

Refer to Figure 3-3 if you're unsure of the names of any of the systems or where they're located. In this discussion, I'm assuming you already have a set of batteries in the Spybot and the controller. If you're unsure about batteries, refer to the "Choosing Among Battery Alternatives" section in Chapter 4.

In many of the descriptions in this section, I'll talk about the front of the Spybot brick or the back of the controller. To reduce confusion among agents, I need to have a standard way of referring to Spybot parts.

When I'm talking about the *Spybot brick*, pretend you're actually inside the canopy and driving the Spybot. The touch sensor is at the front, the communications port is at the back, and the battery compartment is at the bottom of the brick.

When I'm talking about the *handheld controller*, hold it in your hand and point the red button at the Spybot. The red button is at the front, the mode switch is at the back, and the battery compartment is at the bottom of the controller.

Spybot Brick Pushbuttons

The Spybot brick has two *pushbuttons* on the top of the canopy (see Figure 3-4). The black button is the power on/off switch. You press it once to turn the Spybot on. When the Spybot turns on, it will emit two high beeps and a low beep. Next, the arc lights will indicate the current security level you have reached, and then they will alternate between red and green. If you press the power switch again, the Spybot will emit one beep and turn itself off.

Mission Start/Stop or Run Button   On/Off Button

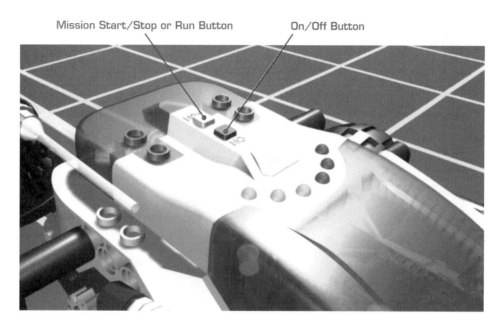

▲ **Figure 3-4.**
The Spybot pushbuttons

The gray Run button is the mission start/stop switch. You press the
button once (after you've turned the Spybot on) to start a mission, and
you press it again to abort the mission.

> **TIP** If you abort a mission, you will affect the points and
> ranking for your Spybot partner. S.M.A.R.T. keeps track of
> the number of times you abort, or give up on a mission, so try to
> complete missions as often as possible! Turning your Spybot off by
> pressing the on/off button when it is trying to complete a mission is
> considered "aborting" a mission and will reduce your agent ranking.

Spybot Indicator Lights

The *indicator lights* (see Figure 3-5) provide valuable feedback from the
Spybot to the agent, such as security levels, the direction to the nearest
controller, and even when a mission timer is about to expire. The lights
operate differently depending on the mission that has been downloaded
to the Spybot.

▲ **Figure 3-5.**
The Spybot indicator lights

Each Spybot has a set of lights that are arranged in a semicircle, or
arc. The video presentation on the Spybotics CD-ROM also calls them
arc lights. Once again, pretend you're sitting in the driver's seat of the
Spybot. The three lights on the left side of the arc are red, and the three
lights on the right side are green.

Your initial security level is 1, so after you turn your Spybot on by
pressing the on/off switch, a single red light will turn on before the arc
lights alternate between red and green. At the highest security level, all
of the red and green lights, as well as the yellow light, will turn on.

NOTE I'm not sure if it's just lucky or by design, but the Spybot arc lights are arranged in the same way as navigation lights on boats and aircraft. The red lights are on the left (port) side of the vehicle, and the green lights are on the right (starboard) side.

The single yellow light is used for different things depending on the mission. In some cases, it tells you that the light sensor has been calibrated correctly. It can also let you know if time is running short or if you are receiving damage.

Refer to the "Mission Summaries" section in Chapter 6 for detailed information on how the indicator lights work in each mission.

Handheld Controller/Beacon Modes

The LEGO Spybotics *handheld controller/beacon* has many interesting features that are not immediately apparent. The most important thing to be aware of is the position of the mode selector switch (see Figure 3-6). This section has special information that will help you get the most out of your Spybot when operating it by yourself or with other agents.

Link Mode   Remote Control Mode   Action Control Mode

▲ **Figure 3-6.**
The Spybot handheld controller/beacon modes: Link, Remote Control, and Action Control

The front of the controller has five pushbuttons and a mode selector switch. The back of the controller has a battery compartment cover as well as a translucent cover that protects the infrared communications channel.

> **CAUTION** The infrared transmitter in the Spybot controller uses the same kind of technology as a television remote control and the range finder in a video camera. When you use your Spybot, be aware of these sources of interference if your Spybot does not follow your commands. Do not cover the clear plastic of the remote with your hands, and try to keep the remote aimed at your Spybot.

To make full use of the Spybotics controller/beacon, you need to be aware of the three *modes* it can be in: Link mode, Remote Control mode, and Action Control mode. The next sections discuss each mode.

Link Mode

Before you start communicating with your Spybot using the handheld controller, you need to establish a secure communications link between the Spybot and the controller. Whether you choose to execute your missions yourself or join forces with other agents, you need to make sure your Spybot responds only to your commands by setting up a secure communications link.

EXERCISE

Setting Up a Secure Communications Link

Your secure communications link, also called a *comms link,* is easy to set up. First, make sure that your Spybot is powered up and ready to link. The two middle indicator lights should alternate between red and green. Then follow these steps:

1. Put the controller in Link mode by making sure the switch is in the middle position.

2. Press one of the channel select buttons: 1, 2, or 3.

3. Put the controller about 6 in. (15 cm) away from the front of the Spybot.

4. Press the Link button (the button with the picture of a key beside it).

You'll hear a series of high-pitched beeps if the link works. If you don't hear the beeps, then move the controller closer to your Spybot and press the Link button again. If you still can't get your comms link set up, you might consider checking the state of your batteries.

Remote Control Mode

Remote Control mode is what you use to drive the Spybot around. Make sure the switch is pointing to the left if you want to be in Remote Control mode. If you recall the discussion on differential drives in

Chapter 2, you understand how the two pushbuttons on each side of the controller work. Table 3-1 shows which buttons (the lighter colored ones) to press to drive your Spybot in the direction you want it to go.

▼ **Table 3-1.**
Remote Control Button Combinations

PUSH THIS BUTTON . . . TO MOVE LIKE THIS					
Forward	Backward	Turn left	Spin left	Turn right	Spin right

By now you're probably wondering what the red button between the direction controls does. The answer is that it depends on the mission you are trying to perform. In the Energy Crisis mission, for example, the red button gives you a shield against damage, but you can only use it six times. In the Gamma Overload mission, the red button sends a command that tells the Spybot to seal off radiation leaks. Refer to Chapter 6 for full details of controller settings for each mission.

Action Control Mode

You use Action Control mode to give extra powers and actions to your LEGO Spybot. In Chapter 7, you'll learn the details of controller settings for different missions. Make sure the switch is pointing to the right if you want to be in Action Control mode. If you look into the clear canopy on the bottom of the controller, you should see a flashing red light to confirm that you're in Action Control mode.

In this mode, your controller also acts as a beacon. The controller sends out a steady stream of infrared energy that the Spybot can use to determine its range and distance. The Command Override mission uses the controller in Action Control mode as a homing beacon. Your Spybot will move toward the controller all by itself!

You can assign each of the buttons a specific command or action. The function you assign to the red button on the controller works the same in both Action Control mode and Remote Control mode.

Infrared Range Finding

If you've read the previous section carefully, you know that you can use the controller as a beacon. This section describes how the Spybot range finder works, and it also has a little experiment you can do to verify your Spybot's range finder.

Infrared (IR) light is a kind of energy that is invisible to humans. It is used in many ways—for example, it is used in television remote controls, camera range finders, and night-vision goggles. Sunlight and even fluorescent lights are also sources of IR light. Because the Spybot range finder and communication systems both use IR light, you must be aware of unwanted sources of IR light. They may cause interference and a loss of control of your Spybot.

You can find the IR detectors that the range finder uses behind the colored canopy on either side of the touch sensor (see Figure 3-7). The range finder works by measuring the intensity of the pulses of IR light from your controller or another Spybot. Each Spybot has three emitters. You can see two of them under the canopy on either side of the indicator lights and another one facing the rear of the Spybot, just above the programming cable connection. On the controller, the emitter is the whitish cylinder on the left side of the clear cover.

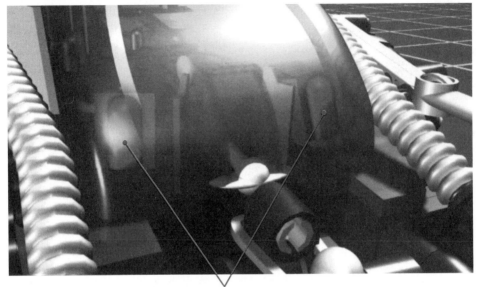

Infrared Range Finders

▲ **Figure 3-7.**
The Spybot infrared range finder

The range finder actually measures two things: the distance to an object that has an emitter (range) and the direction to the object (bearing). Each of the detectors can measure the intensity of IR light that it sees.

When the light source is straight ahead of the Spybot, both detectors "see" the light equally. The computer inside the Spybot figures out that the light is straight ahead. If the light source is to the left of the Spybot, then the left side detector will see more light because of the way the detectors are angled. This helps the Spybot to easily figure out the bearing of the source.

As the source of light moves farther away, the intensity of the light seen by the Spybot zone finder gets less and less. To reduce the complexity of the system, the Spybot defines three zones for objects: Here, There, and Anywhere (see Figure 3-8).

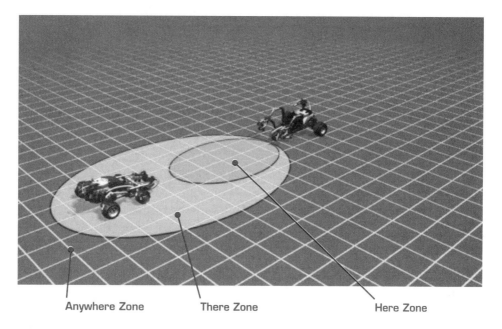

Anywhere Zone   There Zone   Here Zone

▲ **Figure 3-8.**
The Spybot zones: Here, There, and Anywhere

The Here zone is an area within about 15–18 in. (40–50 cm) in front of the Spybot. The There zone is an area within about 72–80 in. (180–200 cm) in front of the Spybot. The Anywhere zone is literally anywhere else.

The exact value of each zone limit depends on many things, such as the angle between the controller and the Spybot, the height of the controller, the level of light in the room, and even battery strength.

EXERCISE

Mapping the Here, There, and Anywhere Zones

You can make a "map" of the Here, There, and Anywhere zones for your Spybot and controller as follows:

1. From the Technical Manual screen, you can download the Spybot Zone Test Program (see the instructions earlier in this chapter for how to download the Zone Test Program).

2. Put your controller into Action Control mode (move the switch to the right) and place the controller in front of the Spybot.

3. If you turn on your Spybot and press its Run button, the indicator lights on the Spybot will show the direction of the controller.

The yellow light will flash quickly when the controller is in the Here zone, and it will flash slowly when the controller is in the There zone. The yellow light turns off and the indicator lights scan when the controller is in the Anywhere zone. As you move the controller around the Spybot, you can get a feeling of where the zone limits are. They should look like the limits shown in Figure 3-8.

The zones are important because you'll use them to program your Spybot to do specific things as you learn more about custom missions in Chapter 7.

Spybot Data Port, Light Sensor, and Laser Port

The Spybot *data port* is the connector at the back of the brick. The *light sensor* and *laser port* are actually buried inside the back of the Spybot. When you need to exchange information between your Spybot and your PC, you must remove the light sensor and laser connectors from the Spybot so you can plug in the data cable. The data cable only fits one way, and it is polarized by using a triangle-shaped pin in the cable that fits into the hole in the Spybot.

Light Sensor Port   Laser Port

▲ **Figure 3-9.**
The Spybot data port, light sensor port, and laser port

After you download a mission to your Spybot, you need to connect the light sensor and laser ports again. It's pretty hard to mix them up because the length of the tubing makes it awkward to connect to the wrong port. If you're unsure, check the symbol just below each port. The light sensor port has a picture of an eye and the laser port has a symbol similar to the universal laser symbol.

 NOTE The universal symbol for hazardous laser light is shown here: Note that the line goes from right to left. The symbol on the Spybot runs from left to right. In any case, *there is no actual laser light emitted from the Spybot,* so it is completely safe.

If you've already assembled your Spybot, you probably know that a length of clear, flexible, and rubbery plastic runs from the port to both the laser port and the light sensor. This is called a *light pipe.* The light pipe that plugs into the light sensor is assembled inside a flexible black hose. The laser port light pipe isn't covered.

A light pipe actually channels light along its length from one end to the other using the same physical principles as fiber-optic cables. You can prove this to yourself by removing the light sensor assembly from your Spybot. Hold one end near one of your eyes, but far enough away to avoid poking yourself. Move the other end toward a light source or a brightly colored object. You will notice that your end of the light pipe shows you what the other end of the light pipe sees.

NOTE This property of light pipes is used in an instrument called an *endoscope.* An endoscope allows a doctor to look inside your body using a bundle of fiber-optic cables without having to risk a large-scale surgery.

Commercial-grade fiber optics are engineered to *close tolerances,* which means that one fiber is almost exactly like the next one in a bundle. It also means they are very expensive. The light pipe used in the Spybots is much less costly than commercial-grade fiber optics because it is made of an inexpensive flexible plastic material that works in much the same way as the glass in a real fiber-optic cable. The light pipes work by taking advantage of *total internal reflection*—but what does that mean?

If you're in a darkened room and you shine a flashlight straight out the window, most of the light passes through the glass to the outside. As you shine the light at more of an angle, eventually you get to the point at which the glass acts as a mirror and reflects almost all of the light back into the room. This is the *angle of total reflection* (see Figure 3-10).

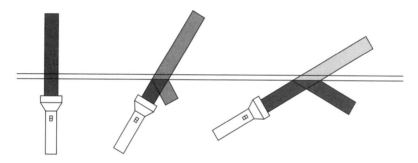

The more you tilt the flashlight, the more light is reflected
and the less light passes through the glass.

If the angle is low enough, the light reflects
within the light pipe and comes out the other end.

▲ **Figure 3-10.**
Total internal reflection in a light pipe

The light pipe works in almost the same way. When light enters one end of the plastic pipe, it reflects along *inside* the pipe and eventually comes out the other end. The laser light pipe works the same way, except in reverse. Instead of moving light from outside into the Spybot, the laser pipe takes light from inside the Spybot and channels it to the laser gun.

> **TIP** Assembling the Technic connector pins (described in Chapter 4) onto the light pipes can be a challenge. The pipe material is very soft and tends to bunch up when forced, which makes it like putting on very tight boots. If you give the connector and pipe a bit of a twist as you assemble them, they go together more easily. Also note that for maximum laser power, you want to make sure that the end of the clear laser gun part is touching the end of the light pipe.

Calibrating the Light Sensor

During the course of some of the missions, you'll be asked to *calibrate* the light sensor by placing the Spybot under the light source and pressing its Run button once. When the yellow indicator light flashes slowly, the Spybot light sensor is calibrated. What you're doing is training the Spybot to recognize a certain light level. When the Spybot light sensor detects the same light level later on in the mission, it will execute its preprogrammed action.

As you become a more accomplished Spybot agent, you'll learn to program the Spybot yourself and take control of its behavior based on readings from the light sensor.

Spybot Default Program

Besides the mission programs—which you'll explore in detail in Chapter 6—there are two other programs you can download to the Spybot. The first is the Zone Test Program, which was described earlier in this chapter. The other is the Default Program, which you download from the Status screen.

EXERCISE

Downloading the Default Program

To download the Default Program, follow these steps:

1. Go to the S.M.A.R.T. home screen and click the gray Spybot Status button.

2. On the Status screen you'll see an entry for Default Program with a triangle beside it. Click the triangle to download the Default Program to your Spybot.

If you don't see a triangle next to the Default Program entry, make sure that your Spybot is turned on and connected to the computer.

You can use the Default Program to demonstrate your Spybot's special skills to other humans you feel would make good Spybotics agents. Here's what it does after you press the Run button:

- The indicator lights turn on in sequence and the Spybot plays a simple tune. You can drive the Spybot around by using a controller in Remote Control mode. Pressing the red button on the controller fires the laser.

- If no other controllers or Spybots are visible, the Spybot spins counterclockwise, makes a chirping sound, and moves forward for some distance. It then moves forward some more while it makes a twisting move and waits.

- If a target is visible (for example, another active Spybot or a controller set to Action Control mode), the Spybot will try to move toward the target. The Spybot stops when the target is on the border of the Here and There zones. If you have a controller set to Action Control mode, you can make the Spybot follow your controller.

- While the Spybot is moving toward the target, the direction of the target is displayed on two of the indicator lights.

- You can change the behavior of the Spybot using the controller buttons. Press button 3 to make the Spybot back away from the controller, or press button 1 to make the Spybot move toward the controller again.

- If the Spybot hits an obstacle, it plays a sound, backs up, and spins clockwise. Then it goes back to its original behavior.

Although you won't get any points or time credits for driving your Spybot in the default programming mode, it's a pretty neat demonstration of some of your Spybot's capabilities.

Summary

In this chapter, you learned about all of the important features of the Spybot electronic systems. The topics covered include the indicator lights, the pushbuttons, the handheld beacon/controller, the infrared zone finder, and the light detector. By familiarizing yourself with these systems, you will gain more knowledge about how your Spybot works.

When it comes down to equally matched Spybots in a mission, the agent with the extra knowledge and the ability to use it will often win. So come back to this chapter whenever you have questions about how the electronic systems work.

SPYBOT CONSTRUCTION AND CARE

The construction of your LEGO Spybot is critical to your success as a S.M.A.R.T. agent. This chapter will give you some useful tips on the following subjects:

- Organizing your workspace
- Accessing the assembly instructions
- Getting parts ready for assembly
- Choosing among battery alternatives
- Storing your Spybot

As a S.M.A.R.T. agent, you're responsible for the operation of your Spybot. By taking a little extra time during construction to learn about the different components, you can gain some of the skills you need to design and build your own Spybots.

Organizing Your Workspace

Your workspace says a lot about the kind of agent you are. A messy environment will lead to lost parts, incorrect assembly, and perhaps even mission failure! A good agent will keep a tidy work area so that his Spybot partner can be put together properly. You can take a few simple steps to make building your Spybot a more enjoyable experience. Another thing to think about is storing your Spybot safely. I talk about suitable containers later in this chapter.

Because the instructions for the LEGO Spybots are all on the CD-ROM that comes with the kit, you'll need to set up your construction and mission download area in front of the computer screen. Try to clear off a work area about the size of a school desk.

You should also have a good light in your assembly area. If you have a desk lamp, you can use it as a target later during the actual missions. Some of the Spybot parts are easy to mix up, and keeping your workspace well lit will help you to see those parts clearly. This makes it less likely that you'll need to take your Spybot apart to find a "missing" part later.

If you're ready to start assembling your Spybot, then proceed to the next section, which is a review of the assembly instructions.

Accessing the Assembly Instructions

The assembly instructions for all the different Spybots are contained on the CD-ROM that is supplied with the kit. This gives you an opportunity to review the other available Spybots for future mission partners.

I assume that you've already installed the Spybotics software onto your computer—if you haven't, go back to Chapter 1 and read the section on software installation and setup. After you launch the Spybotics program for the first time, you'll be shown a short video to introduce the Spybots. Anytime after that, you can just press the spacebar on your computer keyboard to skip the video when you launch the program.

After the video finishes, you'll be presented with the main screen for S.M.A.R.T. Headquarters (see Figure 4-1).

▲ Figure 4-1.
S.M.A.R.T. Headquarters main screen

Simply click the Spybot that is the same as yours to access the construction files. The first phase of preparing your Spybot is the construction, and the second phase is getting your Spybot to communicate with your computer.

Assembly Instruction Tips and Tricks

You need to be aware of a few things when you build your Spybot from the on-screen instructions. The interface for the building instructions is easy to use, but there are a few tricks that will help your progress (see Figure 4-2):

- To go forward or backward one step at a time, use the Next or Previous button.

- When a video is available for an assembly step, the Play button turns green. The Play button is normally grayed out.

- If some of the assembly steps are hidden from view, then use the Hide Parts List button to get rid of the parts list for that step.

- To quickly go to a specific step, use the Slide To Step control.

Parts List for Current Step

Back to S.M.A.R.T. Headquarters

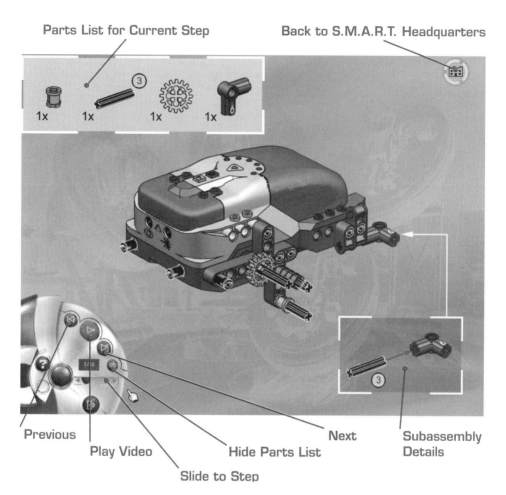

Previous

Play Video

Slide to Step

Hide Parts List

Next

Subassembly Details

▲ **Figure 4-2.**
Spybot assembly instruction control screen

Parts Lists and Diagrams

Along the top of the assembly screen, you'll see the parts list required for each step. It's a very good idea to find all of the parts required for a step before you start assembling them onto your Spybot. That way, you can be sure that you haven't missed anything important. If the parts list is in the way of the diagram, you can always move it or hide it.

The parts list uses a European convention of following the number of parts required with an "x". I automatically translate the "x" as "times" when I read the instruction list. Another thing to watch out for is numbers enclosed in circles. These are used to indicate the length of an axle. You can find an axle length diagram on the instruction sheet that comes with your Spybot. You saved the instruction sheet, right?

If a particular assembly step is unclear, there may also be a subassembly diagram, as you can see in the lower right-hand corner of Figure 4-2. Build all of the subassemblies first and attach them as indicated. Then add all the leftover bits from the parts list to the Spybot.

Remember that the assembly instructions are there to help you ensure that your robot partner is built to specification. Although it is tempting to rush through the construction of your Spybot, try and take a few minutes to figure out *why* each part is used.

NOTE One of the best race car designers of all time was Harry Miller. He worked with Fred Offenhauser and Leo Goossen to produce cars that won the Indy 500 race 15 times between 1921 and 1941. Visit http://www.milleroffy.com for more information on this engineering legend.

One of Harry Miller's most famous mottos was "Simplify, then add lightness." This approach is used in many parts of the Spybots. To keep costs and weight down, LEGO Spybots use only the minimum number of pieces to provide each function.

As you build your Spybot, notice how the parts combine to create the entire machine, one step at a time. Use your powers of observation and deduction to keep track of features that you would like to use on your own customized Spybot. You might also want to browse through the building instructions for the other Spybots for construction hints.

Getting Parts Ready for Assembly

The Spybot kits come in assorted bags. It's a good idea to put all the parts into a container such as a margarine tub that will keep the small bits from rolling off your desk. Just opening the bags and spilling the contents onto your desk is an invitation for trouble.

If this is your first time building an advanced LEGO kit, you'll want to review this section closely. It has some important information on the many different types of pegs that you're going to find in the Spybot kits. Most of the other parts in the Spybot kits are easy to identify, but you need to pay close attention to the pegs.

Identifying Technic Pegs

The main difference between the traditional LEGO lines and the more advanced Technic line is the way you connect parts (see Figure 4-3). Traditional LEGO parts stack on top of each other to build stable structures. The basic LEGO Technic beams have holes through their sides that allow you to connect them using pegs or pins. This opens up a lot of interesting possibilities.

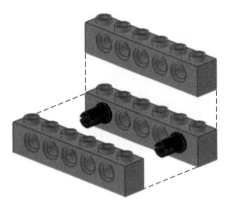

Traditional Connection   Technic Peg Connection

▲ **Figure 4-3.**
Traditional vs. Technic LEGO connections

Depending on what parts you're connecting and how you want them to operate, you'll use one of the many different peg parts available. Knowing what kinds of pegs are available will improve your skills as a Spybot agent and constructor.

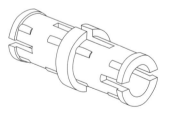

Technic Peg with Friction

This is the simplest and most common Technic peg. This kind of peg is always molded in black plastic. It allows you to connect two beams together firmly. This kind of peg is useful when you need to have a strong assembly that doesn't move much. If you need to connect two beams together to make a wider one, then one peg at each end of the beam will do the job. If you look at the side of your Spybot brick, you'll see a number of holes that you can use for these pegs (step 2/58 of Snaptrax).

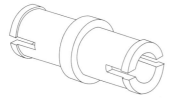

Technic Peg

When you want to connect two beams together so that they rotate or pivot easily, you'll use this part (step 40/58 of Snaptrax). It's always molded in a light gray plastic, which makes it easy to tell it apart from the Technic peg with friction. If you look closely at the two types of pegs, you'll see that this peg is missing the little ridges that give the friction peg its sticking power.

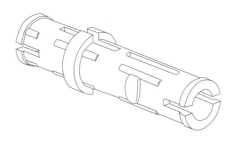

Double Technic Peg with Friction

Sometimes, you'll need to connect three beams together across their sides or connect a beam and some other parts to a Spybot (step 9/58 of Snaptrax). This is where the double peg with friction comes in handy. It allows you to connect more beams without using up more of the holes in the beams. The resulting structure is lighter and stronger than if you used only the standard pins (see Figure 4-4).

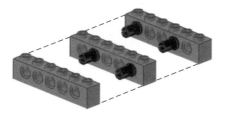

Single Peg Construction

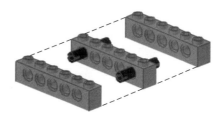

Double Peg Construction

▲ Figure 4-4.
Single vs. double peg construction

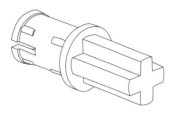

Technic Pegs with Axle

If you've already assembled your Spybot, or if you look closely at some of the beams, you'll see that some of the beams have only peg holes, whereas others have axle holes. Sometimes it's useful to be able to connect beams using the axle holes. Another class of parts that has axle holes is gears, and you can easily connect them to beams using these special pegs.

As you might expect, these two types of pegs come in different colors. The black pegs have friction (step 9/58 of Gigamesh) and are used to provide a more solid connection than the gray ones without friction (step 10/96 of Shadowstrike).

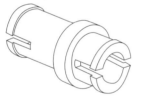

Technic Peg with Half Peg

You use the peg with half peg to connect a full-width element with a half-width part such as a beam. These thinner beams let you make more compact connections to reduce the size and weight of Spybot systems. These parts are always molded in a medium-gray plastic to make them easy to identify when they are mixed in with other parts.

These parts aren't used in the Spybots except to connect the light pipes to the laser and light detector ports (step 43/96 of Shadowstrike).

Technic Peg with Stud

You use the peg with stud (a *stud* is another name for the buttons on the top of a LEGO element) to provide traditional LEGO connection points on the sides of beams. This makes them useful for attaching trim parts onto models. The Spybot models use them mainly in the touch sensor actuator mechanism (step 57/98 of Technojaw). These pegs are molded in light gray plastic. Depending on your computer's display system, these parts are very easy to mistake for the Technic peg with half peg.

Technic Peg Summary

As you can see, a skilled Spybotics robot designer must have a very good understanding of the different types of pegs that are available. Learn to identify the pegs and how they are used in each of the standard Spybots so that you can modify and apply the same techniques in your own creations.

Choosing Among Battery Alternatives

As you and your Spybot partner progress in mastering the different challenges placed before you, eventually your Spybot and controller will need new batteries. This is a good time to think about the different types of batteries that you can use to power your Spybot.

Batteries for the Controller

The handheld controller/beacon uses three AAA (LR-03) batteries. The typical design life of standard alkaline batteries in the controller in Action Control mode is 2 to 3 weeks! The batteries should last for many months of typical usage in Remote Control mode. That's a lot of mission time. It's probably not worth getting anything but standard alkaline batteries for the controller. You won't change them frequently enough to justify any expense or hassle.

> **TIP** Please remember to switch your controller out of Action Control mode and into Remote Control mode when you're not using it! If you leave the controller in Link mode, then an accidental button press might switch to a different communications channel, and your Spybot won't respond to your commands until you set up a new communications link. See the "Handheld Controller/Beacon Modes" section in Chapter 3 for more information on controller modes.

Batteries for the Spybot Brick

The life span of the AA (LR-06) alkaline batteries in the Spybot brick is difficult to estimate. Depending on how carefully you put together your Spybot, and even what kind of Spybot you have, your brick's battery life will vary considerably. For example, the Snaptrax has quite a bit more friction to overcome when it turns than the Technojaw. This is because the tracks must skid along the floor in order to turn the robot.

Another factor that affects Spybot battery life is the quality of the assembly. If the moving parts such as axles or gears are too tight, then the motors will use more power to overcome the friction.

Considering that the batteries in the brick must also power the internal microprocessor, light sensor, laser beam, radar system, and touch sensor, it's not surprising that they don't last as long as those in the controller.

Comparison of Battery Types

There are four types of batteries you might consider using in your Spybots (see Table 4-1). By carefully checking your requirements, you might be able to save some money by using the same batteries for other electronic equipment.

▼ **Table 4-1.**
Battery Type Comparison

BATTERY TYPE	PROS	CONS
Standard alkaline	Easily available Good life	Expensive if replaced frequently
Rechargeable alkaline	Decent life Relatively inexpensive	Limit of about 25 charge cycles Requires special charger
Rechargeable NiCad	Medium price	Medium battery life
Rechargeable NiMH	Easily available Good life	Relatively expensive Self-discharge (i.e., they lose their charge when they aren't being used)

Although LEGO recommends the use of standard alkaline batteries in the Spybots, I have had good success with both the rechargeable alkaline and NiMH batteries. Each of these battery types needs its own charger. Never put a rechargeable battery in the wrong kind of charger. You can use the NiMH batteries in many other electronic devices, such as a digital camera, a CD player, or a handheld electronic game.

> **TIP** You can get NiMH batteries and chargers from your local camera store or consumer electronics store. You can also ask a senior agent (such as a parent) to visit http://www.ripvan100.com to purchase high-quality industrial batteries and chargers at a good price.

Storing Your Spybot

Storing your LEGO Spybots is a very personal activity, so each agent will have her own way of doing things. Your Spybot will need to be stored in one of two states: assembled and dismantled. It's a good idea to store your Spybot carefully, especially if you take it over to a friend's house or to school to work on missions with other agents. Your Spybot is rugged, but it will appreciate it if you take some care in how you handle it.

CAUTION I probably don't need to tell you this, but it's not a good idea to use your Spybot outdoors. The sun is a powerful infrared light source that can interfere with the remote control. Sand and grit can get into the insides of your Spybot and shorten its life. If you do get the Spybot a bit dirty, just wipe it with a cloth. Never immerse your Spybot in water, or it will be ruined.

So far, the very best boxes I have found for storing Spybots (and any other LEGO Technic parts) are the Plano tackle boxes (http://www.planomolding.com). Similar boxes are also made under the TackleLogic brand name. You can get these boxes at any good discount retailer, hardware store, or sporting goods store.

I generally divide the Spybot kit into two types of parts. One type is "fiddly bits" that get lost easily, and the other is "big bits" that can't get easily sucked up in a vacuum cleaner. Each type of part has different storage needs. When you're looking for storage boxes for small LEGO Technic parts, the experts agree on the following requirements:

- Transparent so you can easily see the contents
- Secure lid so the parts don't spill out
- Movable dividers so you can store different-sized parts

- Full-height dividers so parts don't shift between compartments

- Rounded bottoms to make scooping parts out easier

- Easily available and inexpensive so you can buy more LEGO

The Plano 3500 Tackle Box

For the fiddly bits, I recommend you get one Plano 3500 (see Figure 4-5) box for each Spybot you own. With careful planning, you can get all of the little parts for one Spybot into one of these boxes. You can put pegs, beams, axles, catch blocks, gears, and tubing into their own compartments to make them easier to find when you build your Spybot. There's another reason why I like the 3500, but I'll leave that until later.

▲ **Figure 4-5.**
Plano 3500 tackle box with Spybot fiddly bits

The Plano 1140 Tackle Box

Each of the Spybots has some parts that just won't fit into the Plano 3500 tackle box. These include the handheld controller, the Spybot brick, and the large wheels or tracks that drive the Spybot. You'll want to find a box that's big enough to hold the assembled Spybot and the little tackle box too. The Plano 1140 tackle box (see Figure 4-6) fits the bill perfectly.

▲ **Figure 4-6.**
Plano 1140 tackle box for assembled Spybot

The Plano 1258 Tackle Box

Once you have two or more Spybots, or if you have lots of other LEGO Technic parts that you need to store and transport safely, it's time to move up to the Plano 1258 tackle box (see Figure 4-7). This box has lots of adjustable storage on the top layer, holds up to seven of the 3500 tackle boxes on the bottom layer, and has a generous space for holding partially assembled robots.

▲ **Figure 4-7.**
Plano 1258 tackle box for Spybot experts

Summary

I've covered a lot of basic material in this chapter that can make building and maintaining your Spybot partner easier. From organizing your workspace and identifying some of the trickier Spybot parts to storing and transporting your Spybot, there are plenty of things an agent must know to be effective.

CHAPTER 5

SPYBOTICS AGENT COMMUNICATIONS

In this chapter, you'll learn how to read the mission results out of your LEGO Spybot and how to share your results with other agents around the world on the Internet. I also discuss how to exchange your custom missions with your friends.

This chapter covers the following topics:

- Reading the Spybot Mission Summary screen
- Checking your Spybot's status, including battery strength and your security level
- Calculating your Agent Ranking bars based on mission data
- Registering with S.M.A.R.T. on the Internet

Spybot Mission Summary

As you run each mission, your Spybot keeps track of a number of specific items such as

- The number of times you have attempted a mission
- The number of times you succeeded or failed
- How many points and bonus points you have earned
- Your security level
- The total time your Spybot has spent running missions

Every time you connect your Spybot to your computer and the Spybotics software is running, the Spybot sends any new mission data to the computer (see Figure 5-1). This screen overlay can pop up anytime.

▲ **Figure 5-1.**
The Spybot Mission Summary overlay

In the title bar, you will see the name of the latest mission that the data is for. You will also see the number of attempts, successes, and failures, as well as the mission, bonus, and total points scored since the last upload. You close the screen by clicking the "X" in the upper-right corner.

> ▶ **TIP** It's easy to get confused about what direction the data is flowing when you're talk about "uploading" and "downloading" information. An easy way to think about it is in terms of the size of the two devices that are communicating. Imagine that the bigger or smarter device is in charge of the data. The bigger device *downloads* when it sends data and it *uploads* when it receives data.
>
> In your case, the computer is the smart device, so when you send data to the Spybot, you're downloading to it. When the Spybot sends data to the computer, it's uploading information to you.

The programmers at S.M.A.R.T. force your Spybot to communicate its status to your computer every time the Spybot connects to the computer. This is to ensure you don't lose your points by accident when you download a new mission to the Spybot.

Spybot Status Screen

In Chapter 3 you had a quick look at the Spybot Status screen. In this section, you're going to examine it closely and find out all kinds of interesting things about your Spybot partner (see Figure 5-2).

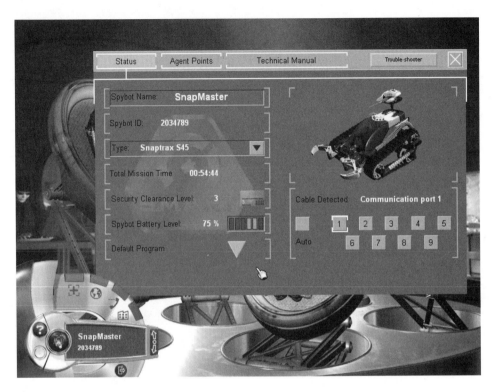

▲ **Figure 5-2.**
The Spybot Status screen details

Table 5-1 outlines the Status screen fields.

▼ **Table 5·1.**
Status Screen Fields

FIELD	DESCRIPTION
Spybot Name	You can change the name of your Spybot at any time and you won't lose any points that you earned. The name is limited to 15 characters.
Spybot ID	Each Spybot gets a unique digital ID when it comes from the factory, and this ID can't be changed. The ID is used by the Spybotics software to keep track of points and progress. You can also use the ID to confirm which Spybot is yours if you accidentally switch Spybots with another agent.
Type	You set the type of your Spybot in this field. You can only choose one of the four basic types: Snaptrax, Gigamesh, Shadowstrike, or Technojaw. Be careful when you change this setting—one of the Spybot types will cause the others to run backward! I'll leave this as a secret for you to figure out.
Total Mission Time	This is the total time that your Spybot has spent running missions with you. The clock does not run if you just leave your Spybot powered on. You can only accumulate mission time by running missions.
Security Clearance Level	Your current security level is displayed by the indicator lights every time you power up the Spybot. You can also check the level out on this status display.
Spybot Battery Level	The current battery level is shown here as a percentage of a fully charged alkaline battery. If you choose to use rechargeable NiCad or NiMH batteries, they will show between 75% and 80% even when fully charged.
Default Program	This is where you load the default software into the Spybot. Chapter 3 discusses its operation in more detail.

Spybot to Computer Link Connection Status

There are two other areas of interest on the Spybot Status screen. The first is the picture of the Spybot that changes to match the type of Spybot that is actually connected to the computer. Double-check the picture with your Spybot to make sure they match. If your Spybot is behaving strangely, maybe another agent has changed the type to one with a different gear train setup.

Just below the picture of the Spybot is the current status of the cable connection. Usually, you only need to set this up once. If your Spybot is not being detected, you should check this part of the Status screen. If there is no cable detected and you are sure that it is actually plugged in, click the Auto button to rescan the serial ports. If the cable is detected, there will be a white outline around the serial port that it is plugged into.

Spybotics Agent Points Screen

If you want an overview of how you and your Spybot are doing as a team, just click the Agent Points menu entry to bring up the Agents Points screen (see Figure 5-3). This screen tells you a lot about how you and your partner are doing.

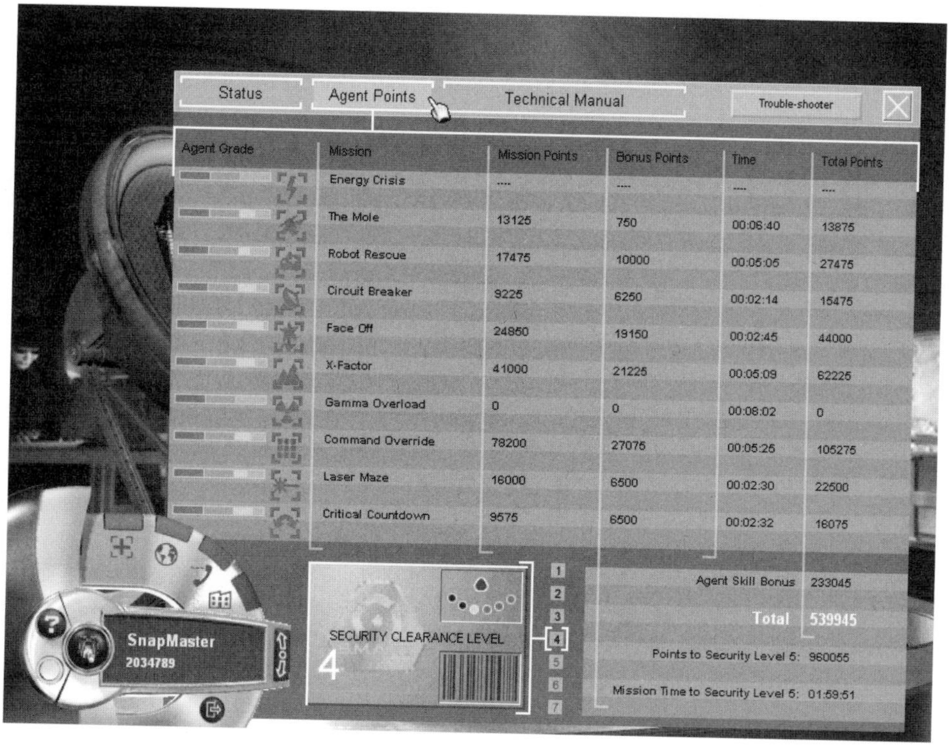

▲ **Figure 5-3.**
The Spybotics Agent Points screen

Each of the ten built-in missions is displayed along with the current
Agent Grade ranking for that mission. There is also information on the
mission points, bonus points, and time spent on each of the missions, as
well as the total points and your Agent Skill bonus.

Agent Grade Ranking

The Agent Grade ranking is displayed as a bar graph that makes it easy to see how well you're doing at particular missions. There's a bit of simple math involved in figuring out what the bars mean, but if you can figure out your favorite sports team's win/loss ratio, you can handle this.

The Spybotics software keeps track of lots of information on your missions. The Agent Grades depend on five different statistics for each mission:

- Wins
- Losses
- Attempts
- Points
- Max points available for the mission

Agent Ranking Bars

The three Agent Ranking bars represent a range of values from 0 to 100. The longer the bar, the closer your score is to a perfect 100. The bar values are calculated as shown in Table 5-2.

▼ **Table 5-2.**
Spybotics Agent Grade Bar Calculations

BAR COLOR	BAR NAME	FORMULA
Red	Completion Rate	((Wins + Losses) / Attempts) * 100
COMMENT: Only displayed if you have *attempted* the mission more than two times		
Orange	Win Rate	(Wins / (Wins + Losses)) * 100
COMMENT: Only displayed if you have *completed* the mission more than two times		
Yellow	Points Rate	((Points / Wins) / Max Points) * 100
COMMENT: Only displayed if you have *beaten* the mission at least once		

EXERCISE

Calculating Your Agent Ranking

Calculating your agent ranking by yourself might be an interesting exercise for some agents. Here's how to do it.

The information you need is only available from the Spybotics Web site—that's where S.M.A.R.T. keeps track of your Spybot's progress. You'll need to read the section on registering with S.M.A.R.T. in the "Registering and Connecting with S.M.A.R.T." section in this chapter first. You'll also need a calculator to help you with the division and multiplication.

When you connect with S.M.A.R.T., you will get a complete summary of each Spybotics mission, including attempts, wins, and losses. For this example, let's use this information:

Attempts = 30

Wins = 9

Losses = 17

Points = 126,500

Max Points= 18,750

The red bar indicates the completion rate. Abandoning a mission is neither a win nor a loss, so this number tells you how much of a "quitter" you are. Use the Completion Rate formula from Table 5-2 like this:

((9 Wins + 17 Losses) / 30 Attempts) * 100 = 86.6

The orange bar indicates the win rate. Abandoned missions are not used in the calculation, so wins against losses gives an indication of how good you are at beating the mission. Use the Win Rate formula from Table 5-2 like this:

(9 Wins / (9 Wins + 17 Losses)) * 100 = 52.9

The yellow bar indicates the points rate. Only winning missions count toward your total points, and harder missions get you those points faster. This bar measures the difficulty level of the missions you beat. Use the Points Rate formula from Table 5-2 like this:

$$((126{,}500 \text{ Points} / 9 \text{ Wins}) / 18{,}750 \text{ Max Points}) * 100 = 74.9$$

Other Agent Points Information

You might wonder why there is a difference between mission attempts and mission completions in the S.M.A.R.T. Agent Ranking system. This is due to a fundamental principle within S.M.A.R.T.:

A winner never quits, and a quitter never wins.

The system adds 1 to the number of attempts every time you press the Run button, but it only increments the wins or losses if you finish the mission. Attempting missions without completing them will hurt your Completion Rate bar, whereas beating missions improves the value of all the bars.

When you compare yourself with other agents, make sure you look closely at the yellow Points Rate bar. It's not too hard to get good Completion Rate and Win Rate bars. The points rate is a measure of the true difficulty of the missions that you are attempting. The (Points/Wins) value gives you the average number of points you earn every time you do the mission. Dividing the result by the Max Points value gives you a bigger number for more difficult mission settings. You can check the Max Points value by setting each mission to maximum difficulty.

At the bottom of the Agent Points screen is an indication of the mission time and points you'll need to log to get to the next security clearance level.

If you are lucky enough to have more than one Spybot partner to choose from, each one will have its own summary screen. Remember the Spybot ID in the Status screen? This is used to keep track of individual statistics, so even if you have three Shadowstrike Spybots, each has its own summary screen.

Registering and Connecting with S.M.A.R.T.

Eventually, you'll want to rank yourself against other agents around the world. Just click the telephone icon on the communicator to contact S.M.A.R.T. via the Internet. You will be asked to make sure your Spybot is connected to your computer and switched on before you are allowed to contact S.M.A.R.T.

> **CAUTION** It's a good idea to get help form a more senior agent (such as a parent or older sibling) the first time you contact S.M.A.R.T. At no time are you asked to give personal information that could be used to identify you by name or where you live. Never give out private data about yourself on the Internet—if you're unsure, ask a grownup for help.

Just before connecting to S.M.A.R.T., the Spybotics program reminds you that you must have access to the Internet. Your default Web browser (Microsoft Internet Explorer for most of you) will open up.

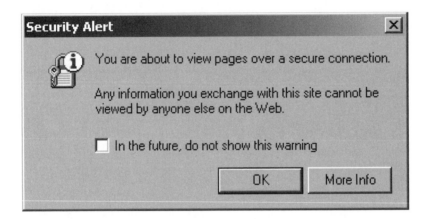

> **TIP** Depending on the security settings you or other agents in
> your home have chosen for the Internet, you may get a
> warning like this when you contact S.M.A.R.T.:

Check with a more senior agent if you are unsure whether or not
to proceed. S.M.A.R.T. has set up its Web site to use a secure
connection to keep your identity private. If you read the warning text,
it just says, *"Any information you exchange with this site cannot be
viewed by anyone else on the Web."* This is good, so click OK to
proceed.

If you get this warning when you log in, you will get the same warning
after a successful login when you proceed to the main Web site.

If this is your first time on the S.M.A.R.T. Web site, you'll need to
register before you can log in (see Figure 5-4). Just click the "SMART
Network" text at the bottom of this screen to register. It's a bit hard to
tell this text is a link, but it is. You'll be asked to fill in some information
so that S.M.A.R.T. can verify your identity later. The only personal information is your birth date, the country you live in, and your gender, which
can't be used to identify where you live or exactly who you are, so it's
safe to fill in these fields.

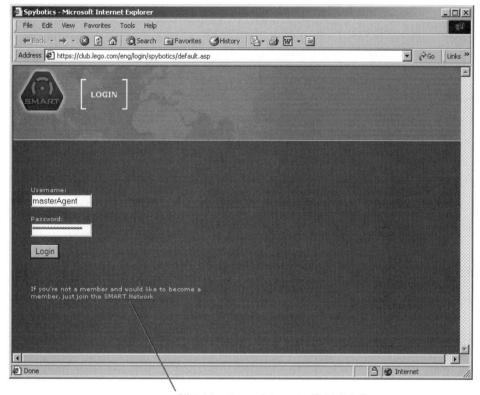

Click here to register with S.M.A.R.T.

▲ **Figure 5-4.**
The S.M.A.R.T. Web site registration screen

If you forget to fill in a field, the Web site will tell you what went wrong. If you're younger than 13 years old, you'll be asked to provide the e-mail address of a parent.

Once you're registered, you can log into S.M.A.R.T. at any time. You'll be given a choice of checking your own status or finding out what specific areas of the globe need your help. Click the Agent Status area to bring up your individual summary. Along the right side, you'll see little circles that you can click to compare yourself with the nine agents that contacted S.M.A.R.T. most recently.

Scroll down the screen to compare yourself with another agent on a mission-by-mission basis. At the very bottom of the screen, you'll find a search area where you can find the robots of other agents if you know their nicknames. If you do an advanced search, you can find other Spybots by their type or their security level.

Let's say you want to find other Gigamesh Spybots that are at Security Clearance 4. Fill in the search fields as shown in Figure 5-5 and click the Go button. You may need to scroll down to actually see the results. The results will show the names of other Spybots that match your search criteria. You can then click the names to bring up the mission summaries for that Spybot to compare with your own.

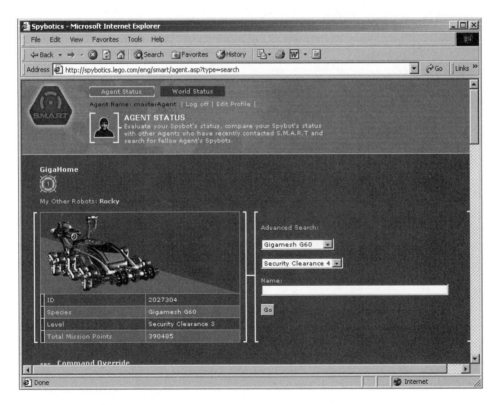

▲ **Figure 5-5.**
The S.M.A.R.T. Spybot Advanced Search screen

If you click the World Status button, a screen summarizing the state of the Spybotics world is shown (see Figure 5-6). S.M.A.R.T. collects Spybot mission results from other agents like you and keeps track of which mission areas need help. The red areas are in the most danger, so get over to that area with your Spybot partner and help out if you can!

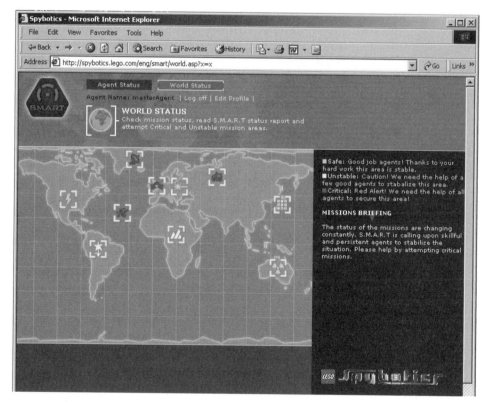

▲ **Figure 5-6.**
The S.M.A.R.T. World Status screen

Exchanging Custom Missions with Friends

Although I don't talk about custom missions until Chapter 7, I might as well cover how to exchange missions with friends over the Internet. The information for each of the Special Operations missions you create is kept in its own file. The location of the file is C:\Program Files\ LEGO Software\Products\Spybotics\Bookmarks\SpecialOperations\MissionName.lif, where MissionName is the name you have assigned to the Special Operation. Just attach the file to an e-mail and send it to your friend. Make sure your friend puts the file in the same folder on his machine; otherwise, it won't show up when he opens the Special Operations menu.

Summary

You have now reviewed all of the background material that you need to tackle missions with the confidence to succeed. You know all about the Spybot mechanical and electronic systems, and you have carefully built and tested your Spybot. Finally, you have connected your Spybot to your PC and perhaps even communicated with S.M.A.R.T.

Agent, you may proceed to the next chapter to review the detailed mission briefs. Then you can move on to designing your own custom missions.

6

MISSION SELECTION GUIDE

At this point, you are ready to choose a mission that will start you on your way to earning higher security ratings. Be careful, Agent—there is much to learn about each mission that will help you succeed!

This chapter covers the following topics:

- Introducing the mission profiles
- Setting up your mission area for best results
- Understanding the mission setup screens
- Downloading missions to your Spybot
- Viewing detailed mission summaries

Basic Mission Profiles

Each of the mission descriptions in this chapter follows an easy-to-understand format that describes the mission brief, explains how to set up the mission area, and even offers some handy hints for better results. Some of the missions are designed for one Spybot and one agent, whereas others are designed for up to two Spybots and two agents. Table 6-1 summarizes the basic mission profiles.

▼ **Table 6-1.**
Spybotics Mission Summaries

MISSION NAME	NUMBER OF AGENTS	NUMBER OF SPYBOTS	MISSION SUMMARY
Circuit Breaker	1	2	Keep the Earth's climate stable by intercepting a Hackerbot that is trying to take over the weather-control satellites.
Command Override	1	1	Crack the code to free your Spybot from a control virus planted by enemy agents.
Critical Countdown	1+	2+	Pass a binary leech to another Spybot before it sucks all your Spybot's energy.
Energy Crisis	1	1	Save the electrical grid by dodging impulse generators and getting your Spybot to the override switch.
Face Off	2	2	Hypnotize an enemy Spybot and return it to your home base so that the secret Hypnosium serum stays in good hands.
Gamma Overload	1	1	Fix a number of radiation leaks in the outer wall of the S.M.A.R.T. test reactor before it melts down.
Laser Maze	1	1	Escape from an old factory by dodging laser beams that only your Spybot can see.

Continued

Continued

MISSION NAME	NUMBER OF AGENTS	NUMBER OF SPYBOTS	MISSION SUMMARY
Robot Rescue	1	2	Rescue a stranded Spybot by sharing your Spybot's energy with it.
The Mole	1+	2	Intercept a data stream being sent by a mole agent and restore the data to the S.M.A.R.T. servers.
X-Factor	2	1	Get as many of the X-Factor fire gems as you can before the mine elevator comes to take you to the surface.

Mission Area Setup Requirements

In many of the missions, you'll require the same types of accessories to set the scene. Don't worry, Agent, you probably already have all the things you'll need to make your missions both fun and challenging. The following tips will help you make the most of your valuable mission time.

Computer Area

Your computer area should be set up with the Spybotics control cable in an easy-to-access place. You need to be able to download the mission to the Spybot and upload the results back to your computer.

If you are using a laptop computer, put it up out of harm's way on a desk. Resist the temptation to place it in the mission area or on a nearby chair or other flat area at sitting level. You or one of your coagents may get caught up in the game and end up stepping or sitting on your only link back to S.M.A.R.T. Headquarters. The most likely form of damage is a cracked display screen, which generally costs as much to fix as a new computer. This is not a good way to end a mission.

Mission Area

The area you choose makes a big difference in the quality of your mission. You should find a room with a clear area of about 6–12 ft. (2–4 m) in size on each side. A smaller area will make it difficult to move your Spybot, especially if other agents are working in the same area. You can have other sectors leading off from the main play area to increase the difficulty of the mission.

The floor should be fairly clean and free of dust, sand, animal hair, and loose fibers. These tend to get into the gears and motors of your Spybot and will increase the friction in the mechanisms, which in turn reduces battery life.

A hard floor surface such as wood, tile, or linoleum is best. If you must have the mission area on a carpet, then aim for one with short, dense fibers. Deep carpeting has a surprising amount of friction and will severely reduce your Spybot's battery life. If you actually still have shag carpeting in the house, ask your parents to replace it. It's probably getting old anyways.

You should also choose a room that has even lighting, or one without some areas that are much brighter than others. The reason for this is that bright lights can confuse your Spybot and make it difficult to control.

Obstacles

Many of the mission briefs suggest using common objects as additional obstacles. These might include stacks of books, chairs, and even houseplants. A good obstacle should be heavy enough to trigger the Spybot bumper without taking up too much of the valuable playing area.

I like to use full soft-drink cans or plastic bottles for obstacles. It is very easy to make a paper sleeve for the can or bottle that you can decorate to resemble the types of obstacles you will find in the mission area.

Another thing you can do to set up a real challenge is to make ramps and catwalks out of cardboard or foam-core board. Don't make them so high that your Spybot is damaged if it falls off. A height of 6 in. (15 cm) is plenty to make things interesting. You can decorate and color your ramps to make the missions more realistic.

Remember to ask for assistance and clearance from a more senior agent (such as a parent) if you need help with cutting or gluing your obstacles.

Light Sensor Goal Areas

Many of the Spybotics missions use lights to represent goal areas or places where your Spybot can get more energy. When you use multiple lamps, it is important that they all have about the same light output so that the Spybot can recognize them properly. Agent, I assume that you know you need to turn the light on so that the Spybot can recognize it!

Just about any small table lamp or desk lamp will do—just make sure that the actual light is about 3 ft. (1 m) from the floor. If you don't have a suitable light, you can also try a flashlight, but be prepared to invest in batteries! Some of the newer flashlights have very bright light-emitting diodes (LEDs) that emit a nice, white light—this kind of flashlight is a perfect addition to your collection of regular spy tools.

Understanding the Mission Screens

You enter the Mission Map screen (see Figure 6-1) from the S.M.A.R.T. home screen or any other screen that displays the communicator by clicking the globe icon. The Mission Map is your entry point to the different missions.

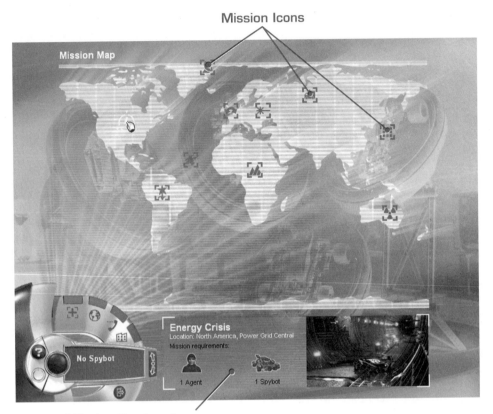

▲ **Figure 6-1.**
S.M.A.R.T. Mission Map screen

As you move the mouse over each of the red "hot spot" icons on the map, a summary of the mission will appear on the lower half of the screen. You will hear your mission supervisor at S.M.A.R.T. describe what the mission is about and you will see some pictures as well. Click the hot spot icon to enter the mission.

TIP If you click the question mark (?) button on the communicator area in the lower-left part of the screen, an audible help system is activated. As you move the move the mouse over icons or other areas of interest, your will hear a description of what the mouse is pointing at. This works anywhere in the Spybotics world, not just in the mission areas.

Each of the individual mission screens (see Figure 6-2 for the Energy Crisis Mission Board screen) has a column of icons along the left side and another icon in the middle of the lower part of the screen. As you move the mouse over each of the icons, you will see a text description of what the icon represents.

▲ **Figure 6-2.**
Energy Crisis Mission Board screen

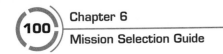

The Mission Board Screen

When you enter a mission, the Mission Board (see Figure 6-2) is the default screen that you are presented with. This is where you can see the rank of your LEGO Spybot against the top ten attempts of any other Spybot that has connected to your computer.

As you complete more runs of the same mission and improve your score, your ranking and security clearance will increase accordingly.

The Mission Brief Screen

Clicking the Mission Brief button will play a short video that explains the "back story" of the mission you have chosen to run (see Figure 6-3). You will learn

- Where the situation is

- The problem you and your Spybot are facing

- The steps you must follow to succeed

- Things to watch out for that will make the mission more difficult

The Pause button in the upper-right corner of the screen allows you to stop the video to take notes. When the video completes, the Pause button becomes a Run button that you can click to play the video again. If you want to stop the video and set it back to the beginning, just click the mouse anywhere on the video area.

Pause Button

▲ Figure 6-3.
Energy Crisis Mission Brief screen

TIP You can also use keyboard shortcuts to control the mission brief videos at any time. The spacebar acts as a pause/run toggle switch. The Esc key stops the video and rewinds it back to the beginning.

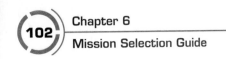

The Mission Set-Up Screen

Clicking the Mission Set-up button will bring up a screen that describes how to set up your Spybot and mission area (see Figure 6-4). There are a number of icons along the left side of the screen and matching buttons across the top. Click the buttons to see a video for each part of the Mission Set-up guide.

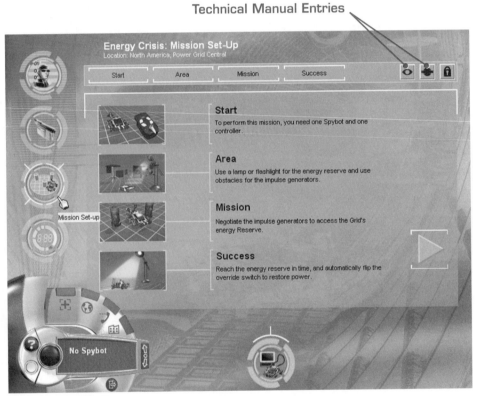

▲ **Figure 6-4.**
Energy Crisis Mission Set-Up screen

If you click the Play icon on the right side of the screen, you will see the Start, Area, Mission, and Success videos play in sequence. There is no pause control for these videos. Clicking in the video area or pressing the spacebar or the Esc key will stop the video immediately.

> **TIP** Along the upper-right part of the Mission Set-Up screen, you will see icons that represent entries in the Technical Manual that you might need to review to understand the mission. In Figure 6-4, you can see icons for the light sensor, the touch sensor, and the comms link. Click each to see its video description from the Technical Manual.

Pay close attention to the Mission and Success videos because they have vital information that will help you understand what to do. Each of the mission summaries in this chapter will go over the information in detail so you can reference it without going to the computer.

Table 6-2 presents a short description of each video.

▼ **Table 6-2.**
Mission Set-Up Videos

VIDEO	DESCRIPTION
Start	This video describes what you need to do to set up your controller/beacon and Spybot. The text beside the button reminds you how many Spybots and controllers you need.
Area	This video shows you how to set up the play area. You'll find out if obstacles are needed, where to place a lamp (or lamps) if needed, and other information about the playing field.
Mission	This video will show you how to calibrate the light sensor if needed and how to start the Spybot on its mission, how to interpret the indicator lights, and other information you need to do your job.
Success	This video describes what happens when you reach the goal. You'll also learn about how bonus points are awarded so you can make the most of your time in the field.

The Mission Settings Screen

The Mission Settings screen is where you get to adjust the difficulty of your missions (see Figure 6-5). Each mission has a different group of settings, or parameters, that you can change to make the mission more complicated or easier, depending on how your Spybotics skills are developing.

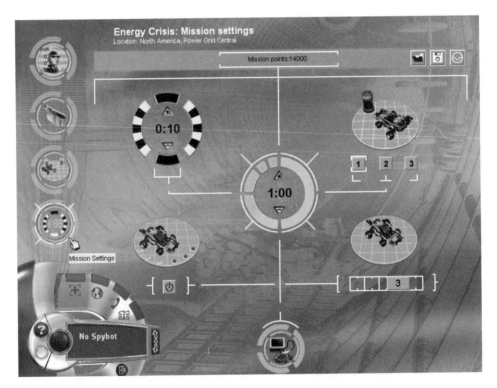

▲ **Figure 6-5.**
Energy Crisis Mission Settings screen

Naturally, increasing the difficulty makes it harder to get bonus points, but it will help to increase your rankings too. You can tell if your adjustments are making the job harder or easier by watching the Mission points area near the top-middle part of the screen. Making the mission harder increases the points and vice versa.

The top-right corner of the Mission Settings screen has three additional icons for saving, loading, and restoring the mission settings (see Table 6-3).

▼ **Table 6-3.**
Mission Settings Icons

ICON	DESCRIPTION
	The Load icon lets you pick any saved settings and bring them into the current mission.
	The Save icon lets you save the current settings so you can load them later.
	The Reset icon puts the mission settings back to the default values in case you make a mistake.

When you load or save your mission settings, you will be asked for some text to identify the settings. You can only use ten characters (including spaces) in the name, and each mission has its own settings area. Try to pick a name that makes it easy to remember what the settings mean. Words like "set1" and "sakjn" are harder to figure out than "easy," "medium," "hard," and "impossible".

The Download Screen

The Download screen is where you can transfer the mission and any custom settings to the Spybot (see Figure 6-6). Click the triangle in the middle of the screen to start the download procedure. Notice that the Spybot picture matches your Spybot type!

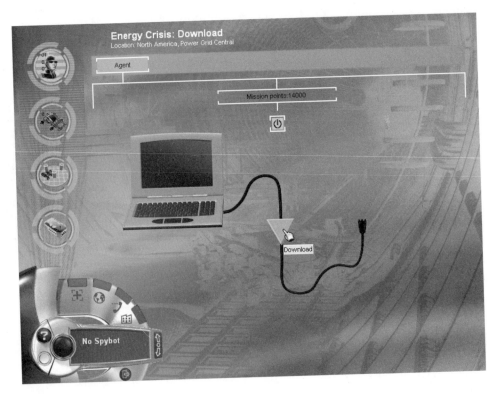

▲ **Figure 6-6.**
Energy Crisis Download screen

The system is intelligent enough to figure out when the download is not working, and it will ask you to make sure your Spybot is connected and switched on if there is a problem. If a retry still does not fix things, you may want to use the Trouble-shooter to help you diagnose the problem.

If you look closely, you'll see the standard on/off icon under the Mission points indicator. You can turn the points-tracking system off to try some "dry runs" to train for your mission without affecting your score. Just remember to turn the points-tracking system back on so that you can advance your security rating!

TIP If you are a very observant agent, you will notice that two of the icons on the left side of the screen have changed. The Mission Brief and Mission Settings icons have been replaced by the Spybot Settings and Controller Settings icons. These are advanced topics that are covered in Chapter 7. For now, I suggest you learn to use your Spybot using the standard settings.

If a mission requires two or more Spybots, an Agent Profile tab and a Rival Profile tab will appear for each additional Spybot across the top. Select the correct profile to download to each of the Spybots in your mission. The software automatically moves to the next tab in multi-Spybot missions, so double-check to make sure the correct profile is uploaded to your Spybot.

Mission Summaries

You're now ready to examine each of the missions. Remember, Agent, you don't need to have your Spybot connected to your computer to review the missions on the CD-ROM. If you use this handbook as a reference, you can learn about missions without having to sit in front of your computer.

Each of the summaries follows the same basic format. You'll first read about the story behind the mission you're about to run. Then there's a section called "Mission Setup" where I explain how to set up the mission area. You only need to do this once. Next, the "Mission Execution" section has the steps you need to follow every time you attempt the mission. It's a complete description of your visual and audio cues from the Spybot, as well as any special functions you can activate with the keys and your controller. Finally, the "Mission Settings" section tells you about the individual parts of the mission that you can control.

A few modes of operation for the indicator lights happen so often that I show them once here to avoid repeating them too often (see Table 6-4).

▼ **Table 6-4.**
Common Spybot Indicators

INDICATOR STATE	SPYBOT STATE	SOUND
	Idle	None

NOTE: The Spybot has just been turned on and the current security level has been displayed. The middle red and green indicator lights are alternating.

|  | Ready for mission | Rising tones followed by ticking once per second |

NOTE: The Run button has been pressed and the Spybot is ready to start the mission. The yellow indicator light is flashing slowly.

|  | Timeout warning | Two warning tones followed by ticking |

NOTE: If the mission time is running short, the yellow indicator light will flash slowly. Other lights may be on depending on the mission.

Continued

Continued

INDICATOR STATE	SPYBOT STATE	SOUND
	Mission success	Spybot theme song

NOTE: If the mission is successful, the arc lights will flash randomly and the Spybot will do a happy dance (similar to what your favorite sports star does when he or she makes a good play).

	Mission failure	Low tones followed by a sputtering, falling tone

NOTE: If the mission fails, all the arc lights will flash at once intermittently and then turn off all together. The Spybot will back up, move from side to side, and then stop (similar to what your favorite sports star does when he or she makes a bad play).

When I describe the "Mission Setup" procedure in each of the summaries, I assume that you've already downloaded the mission to the Spybot. Agent, don't forget to reconnect your light sensor and laser after you've uploaded your mission!

Circuit Breaker: One Agent, Two Spybots

The object of this mission is to keep the Earth's weather systems stable by intercepting a Hackerbot that has compromised the security of a climate control station in the Arctic. The station controls a number of satellites that help to regulate the Earth's climate to prevent global warming. The Hackerbot is trying to control the satellites to warm the Earth up.

Agent, your job is to jam the Hackerbot's signal every time it attempts to realign one of the climate control satellites.

To make things more difficult, you will never be able to tell exactly when the Hackerbot is trying to realign the satellites. Even worse, the Hackerbot has a freeze device that will temporarily disable your Spybot!

Fortunately, the Hackerbot has only a fixed amount of time to realign all of the satellites. If you jam enough of its signals, you'll prevent the polarity reversal and help save the Earth from global warming. If you don't block enough of the signals, you'll fail, the polar ice caps will melt, and the Earth will be covered in water.

Mission Setup

Here's what you'll need to do to set up for this mission:

- Clear a large area of floor space. It's best to try this without obstacles at first. As you get better, add obstacles.

- Download the Hacker profile to the Hackerbot and the Jammer profile to your own Spybot.

- Establish a secure communications link with your controller, and then switch the controller to Remote Control mode.

Mission Execution

Turn on the Spybots and wait for the middle red and green indicator lights to alternate. Press the Run button on each Spybot. When the Spybots are ready, their yellow lights will flash slowly and you will hear a steady series of tones. Table 6-5 shows the indicator states for this mission.

Place your Spybot about 6 in. (15 cm) behind the Hackerbot and activate the touch sensor on either one by pushing the bumper. You will hear the mission start tones from both bots and all of the red and green indicator lights will activate on your Spybot. The lights indicate the time remaining in the mission. The Hackerbot will have all of its lights off to indicate that it has not realigned any satellites.

If your Spybot is within range and can jam the Hackerbot signals, the yellow indicator light will turn on. Pressing the red button on your controller will send a jamming signal to the Hackerbot. If the signal is jammed, the Hackerbot will shudder from side to side for a few seconds and your Spybot will sound some high tones. If you fail to jam the signal, your Spybot sounds a low tone.

When the Hackerbot is trying to realign a satellite, it sounds a warning tone and flashes its indicator lights in sequence. If you are in range, try to jam the Hackerbot's signal by pressing the red button on your controller.

Be careful—if you are within range of the Hackerbot when it fires its freeze gun, you will not be able to move. Your Spybot will make a sound like ice cubes in a glass as long as you are frozen. Your jamming signal will also be disabled if you are frozen, so watch out, Agent!

As the mission progresses, your Spybot will decrease the number of lit indicator lights shown to demonstrate that time is running out for both you and the Hackerbot. When the timer is about to expire, your Spybot makes a clicking sound to indicate a countdown is taking place. If you have prevented the Hackerbot from aligning all of the satellites, then you will succeed. Be very alert, Agent.

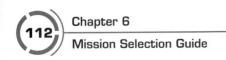
Mission Settings

Table 6-5 contains the settings you'll use during this mission.

▼ **Table 6-5.**
Circuit Breaker Mission Settings

SETTING	DESCRIPTION
	Set the range that you can jam the Hackerbot from by choosing one of two levels. The smaller value means you must be closer to the Hackerbot to jam its signal, which increases the difficulty.
	Set the time required to align the satellites by moving the slider to one of ten levels. The shorter the alignment time, the quicker the Hackerbot is finished realigning and the harder it is for you, the agent, to stop it.
	Set the frequency of satellite alignment attempts by the Hackerbot by choosing one of five levels. Higher values mean the Hackerbot will try to jam signals more often, which increases the danger.
	Set the number of satellites by moving the slider to one of six levels. The fewer the satellites the Hackerbot needs to realign, the harder the mission.
	Set the total mission time in 30-second steps from 1 minute to 10 minutes by clicking the plus (+) arrow or the minus (–) arrow. Shorter mission are easier to complete because the Hackerbot has less time to realign all the satellites.

Command Override: One Agent, One Spybot

The object of this mission is to keep your Spybot from opening a vault containing a large reserve of bullion. No, it's not the little cubes you make soup with—this is *gold bullion*! Enemy agents have captured your Spybot and loaded an encrypted "zombie" program (a kind of virus) and a deactivator that will open the gold vault. Your Spybot no longer responds to commands from the handheld controller.

Agent, your job is to crack the code to disable the zombie program so that the deactivator will not open the vault.

You crack the code by pressing keys on your controller in the correct sequence. The Spybot will tell you how many digits you got right, but not which ones. You must use all of your special reasoning skills to break the code before the enemy agents open the vault using your Spybot.

Mission Setup

Here's what you'll need to do to set up for this mission:

- Clear a large area of floor space. Don't add any obstacles.
- Establish a secure communications link with your controller, and then switch the controller to Action Control mode.
- Place your Spybot at one end of the clear area and place the controller at the other end. The controller represents the gold reserve in the vault.

Mission Execution

Turn on the Spybot and wait for the middle red and green indicator lights to alternate. Press the Run button on the Spybot. When the Spybot is ready, the yellow light will flash slowly and you will hear a steady series of tones.

To start the mission, activate the touch sensor by moving the bumper. You will hear the mission start tones. Quickly go back to your controller and start entering a three-digit code. Your Spybot will advance toward you. Press the controller buttons to enter a digit. Each button causes the Spybot to make a different sound. If the Spybot does not make a sound, then either the button is not valid or you are blocking the controller. In the default mission, only the buttons labeled 1, 2, and 3 are valid.

Remember to release the button and wait a bit after each digit. The controller needs to detect the fact that you have released the button before it can send the next digit to the Spybot. Don't lose your cool, Agent!

After you have entered three digits, your Spybot will make a sound and the indicator lights will show how many digits you got right, but not which ones. Each correct digit in the correct place turns on one green light, and each wrong digit turns on one red light. There are always three lights turned on after a guess.

Table 6-6 is a sample session that shows the indicator lights after each guess.

▼ **Table 6-6.**
Spybot Indicator Lights for Secret Code 323

GUESS	LIGHTS	COMMENT
111		The Spybot tells you that you have three incorrect digits in your guess. This means there are no 1s in the code.
222		The Spybot tells you that you have one digit right and two digits wrong. This means there is at least one 2 in the code. Because you know there are no 1s in the code, there must be one 2 and two 3s in the code.

Continued

Continued

GUESS	LIGHTS	COMMENT
333		The Spybot tells you that you have two digits right and one digit wrong. This confirms your previous guess.
233		The Spybot tells you that you now have only one digit right and two digits wrong. This means you need to rearrange the numbers a bit.
323		Success! You have broken the code just in time!

If your Spybot gets close enough to the controller, the yellow indicator light will start to flash, which means the decode timeout is enabled. The Spybot is getting ready to turn on the deactivator, and you only have a few seconds left to guess the code! If you fail, the Spybot will play a sad sound. Break the code and your Spybot does a happy dance and plays the Spybot theme song.

Mission Settings

Table 6-7 contains the settings you'll use during this mission.

▼ **Table 6-7.**
Command Override Mission Settings

SETTING	DESCRIPTION
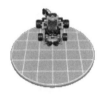	Set the approach speed taken by the rogue Spybot by selecting one of three values. Larger values make the rogue Spybot move faster, which makes it harder for you to crack the code in time.
	Set the value of the highest possible digit in the code. There are always three digits, but each digit can have between two and five different values. The higher the values, the harder it is to crack the code.
	Set the decode timeout in 1-second intervals from 5 seconds to 20 seconds by clicking the arrows. A shorter timeout makes it harder to complete the mission.

Critical Countdown:
One or More Agents, Two or More Spybots

The object of this mission is to get rid of a binary leech that drains your Spybot's energy. You and two or more Spybots have infiltrated a telecom installation to get back some data that has been stolen. Unfortunately, one of the Spybots was detected and was infected with a binary leech. This leech is set to go off at any moment, and when it does, it will suck the energy out of the Spybot it is attached to.

To successfully retrieve the stolen data, you must not have the leech when the timer expires. The only way to get rid of the leech is to give it to another Spybot. If you don't have the leech, then you must avoid the

Spybot that does have the leech. The Spybot with the yellow indicator light on is the one that currently holds the leech.

To pass the leech to another Spybot, you must hit it with your bumper. It's best to hit the other Spybot in the front area to make sure that the leech is transferred. The indicator lights represent the time remaining before the leech is activated, so pay close attention.

Just before the binary leech goes off, the yellow indicator light will flash. You'll have only a few seconds before the leech goes off, so get rid of it by hitting another Spybot. Agent, you must get the stolen data, but you can't do it if you have the leech.

Mission Setup

Here's what you'll need to do to set up for this mission:

- Clear a large area of floor space. If you want an even bigger challenge, add obstacles.

- This mission has one of two personalities for the Spybots. Use the Agent profile if you're going to drive the Spybot using the remote control. If you only have one agent, then program the other Spybot with the Rival profile.

- Establish a secure communications link with each Spybot that will be driven with a controller, and then switch the controller to Action Control mode.

- Place your Spybots about 18 in. (50 cm) apart and facing each other in the middle of the clear area.

Mission Execution

Turn on the Spybot and wait for the middle red and green indicator lights to alternate. Press the Run button on each Spybot. When the Spybots are ready, the yellow lights will flash slowly and you will hear a steady series of tones.

 To start the mission, activate the touch sensor of the Spybot that will have the binary leech first by moving the bumper. You will hear the mission start tones. That Spybot's yellow indicator light will turn on, which means it has the leech.

Drive your Spybot around using the remote control. If one of the Spybots has the Rival personality, you'll see that it will chase you if it has the leech, and it will avoid you if you have the leech. There's a lot more to learn about this automatic behavior in Chapter 7.

Your controller has a number of functions assigned to it in Action Control mode, as shown in Figure 6-7 and Table 6-8. Remember that the red button works the same way in both Action Control mode and Remote Control mode. If one of the Spybots has Rival programming, it will send these commands out at random intervals, so you never really know what to expect!

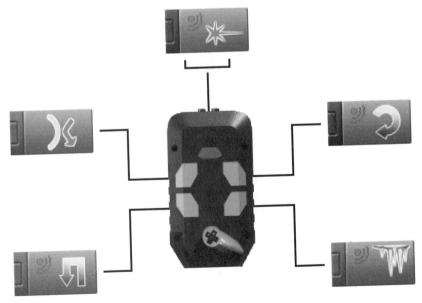

▲ **Figure 6-7.**
Controller settings in the Critical Countdown mission

▼ **Table 6-8.**
Critical Countdown Controller Settings

SETTING	DESCRIPTION
	The Reverse action button reverses the controls of the target Spybot. You'll know that you're affected if everything seems to work in reverse. You can use this on your opponents twice in one mission.
	The Reflect action button sends Laser, Spinner, and ElectroNet actions back to the sender for a short while. You can use this defense twice in a mission.
	The red button sends the Laser fire action, which causes your opponent to stop dead in its tracks and shudder back and forth for a while. You can use this as often as you like.
	The Spinner action button causes the opposing Spybot to spin around for a while. You can use this defense any number of times.
	The Freeze action button will jam your opponent's control signals and freeze it in place for a few seconds. You can use this tactic three times in a mission.

Agent, you must put the controller into Action Control mode for these buttons to have any effect. In Action Control mode, you can't control your Spybot's movement, so you're more vulnerable to getting hit. Switching between Action Control mode and Remote Control mode will take some time, so you'll want to be very careful.

As the mission time winds down, the Spybot that has the leech will begin to flash its yellow indicator light. There is not much time left before the leech goes off and drains its host.

Mission Settings

Table 6-9 contains the settings you'll use during this mission.

▼ **Table 6-9.**
Critical Countdown Mission Settings

SETTING	DESCRIPTION
	Set the timeout warning in 5-second steps from 0 seconds to 30 seconds by clicking the plus (+) arrow or the minus (–) arrow. The shorter the warning time, the harder the mission.
	Set the total mission time in 30-second steps from 30 seconds to 5 minutes by clicking the plus (+) arrow or the minus (–) arrow. Longer missions are easier to complete.
	Set the number of remote-controlled Spybots in this mission from 1 to 3 by clicking the plus (+) arrow or the minus (–) arrow. The total number of Spybots must be at least 2.
	Set the number of automatic Spybots in this mission from 0 to 2 by clicking the plus (+) arrow or the minus (–) arrow. The total number of Spybots must be at least 2.

Energy Crisis: One Agent, One Spybot

The object of this mission is to save the region from a serious power crisis. The *power grid* is the system that distributes electricity from the power station to homes and businesses in the area. A group of enemy agents has sabotaged the grid and now energy is leaking out. Soon there will be no power left!

Agent, your job is to move your Spybot into the power station and get close enough to the energy reserve to flip the override switch to restore power.

But it's not as easy as that. The enemy agents have placed static impulse generators in the access route. If you hit these obstacles, your Spybot will lose energy of its own. Inside the power station are electrical fields that can affect your Spybot at any time. You won't be able to control your Spybot when these fields are active.

If you get within about 15 in. (40 cm) of the energy reserve, your Spybot will automatically flip the override switch and the mission will be over. If time runs out before you complete the mission, you will have failed.

Mission Setup

Here's what you'll need to do to set up for this mission:

- You'll need a clear area of floor space in which to place several obstacles. These obstacles are the impulse generators left by the enemy agents.

- Set up a desk lamp at the edge of the clear space. This is the energy reserve.

- Establish a secure communications link with your controller and then switch the controller to Remote Control mode.

Mission Execution

Turn on your Spybot and wait for the middle red and green indicator lights to alternate. Place your Spybot under the lamp and press the Run button. You will hear a series of four rising tones that indicate the light sensor is calibrating. When the light sensor has finished calibrating, the yellow light will flash slowly and you will hear a steady series of tones.

 Place the Spybot at the opposite end of the mission area and activate the touch sensor by moving the bumper. You will hear the mission start tones and all of the red and green indicator lights will turn on. The number of lit indicator lights tells you how much energy you have left.

Move your Spybot around the mission area using the remote control. Be careful! If you hit one of the impulse generators, the Spybot will back up and then shake from side to side. You will lose one level of energy and some mission time as well.

From time to time, the static electricity fields in the power plant will interrupt communications with your Spybot. This will make it move randomly, send out a series of goofy tones, and flash its lights in a pattern.

 You can activate shields for about 5 seconds by pressing the red button on the controller. The Spybot will sound a rising tone and the yellow indicator light will turn on and stay steady. You can use these shields only six times in a mission and they only work against obstacles. The static generators can still get through the shields. Keep in mind that each time you press the shield button, you reduce the number of shields by one and you get 5 more seconds, even if the shields are already active.

As you get close to the end of the mission timer, the Spybot will sound a warning and the yellow indicator light will flash to signal that you are running out of time. Try to get your Spybot close to the energy reserve before time runs out, Agent!

Mission Settings

Table 6-10 contains the settings you'll use during this mission.

▼ **Table 6-10.**
Energy Crisis Mission Settings

SETTING	DESCRIPTION
	Set the timeout warning in 5-second steps from 0 seconds to 30 seconds by clicking the plus (+) arrow or the minus (–) arrow. The shorter the warning time, the harder the mission.
	Set the impulse generator strength by clicking one of the three levels. Stronger impulse generators increase the difficulty.
	Enable or disable fatigue by clicking the on/off button. If you turn on fatigue, your Spybot will slow down as it loses energy, which makes it harder to win.
	Set the strength of the static interference by moving the slider to one of five levels. More static interference makes it tougher to reach the goal.

Continued

Continued

SETTING	DESCRIPTION
	Set the total mission time in 30-second steps from 30 seconds to 3 minutes by clicking the plus (+) arrow or the minus (−) arrow. Longer missions are easier to complete.

Face Off: Two Agents, Two Spybots

The object of this mission is to hypnotize your opponent and bring him to your base. The S.M.A.R.T. Ecological Research Center has discovered an amazing new serum that comes from the Hypnosium flower. The serum causes instant hypnosis—which can be used for both good and evil purposes—but an enemy agent has stolen some of it. Your job is to use the serum to hypnotize the agent and return to base with him so that S.M.A.R.T. can recover the stolen extract.

You must approach the other agent and sedate him by pressing the red button on your controller. This is like calming down a bear by shooting the bear with a tranquilizer dart. After the agent is sedated, you hypnotize him by touching him with your bumper. He will then follow you back to base. It would be much easier if the other agent stayed hypnotized, but he will eventually wake up and try to hypnotize you, so be careful!

Mission Setup

Here's what you'll need to do to set up for this mission:

- Clear a large area of floor space. To make things really interesting, add some obstacles.

- Establish a secure communications link with each Spybot, and then switch to remote control mode.

● Place your Spybots about 18 in. (50 cm) apart and facing each other in the middle of the clear area.

Mission Execution

Turn on the Spybots and wait for the middle red and green indicator lights to alternate. Press the Run button on each Spybot. When the Spybots are ready, the yellow lights will flash slowly and you will hear a steady series of tones.

To start the mission, activate the touch sensor of either Spybot and return to your home base. You will hear the mission start tones, and all of the Spybot's lights will turn on. The more lights the Spybot has on, the greater the Spybot's resistance to the Hypnosium serum.

Drive your Spybot around using the remote control. When you get close to the front of your opponent, sedate him by pressing the red button on your controller. This will paralyze him so that you can hit him with your bumper. The yellow indicator light will flash if he is sedated.

When you hit the other Spybot with your bumper, he is hypnotized. The indicator lights will flash in sequence from the back to the front of the Spybot. Avoid hitting his bumper with yours, or you may both be hypnotized!

Your controller has a number of functions assigned to it in Action Control mode (see Figure 6-8).

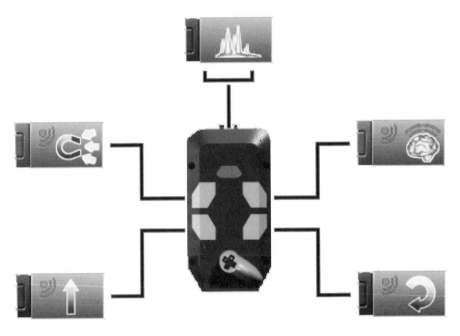

▲ **Figure 6-8.**
Controller settings for the Face Off mission

Remember that the red button works the same in both Action Control mode and Remote Control mode. The functions for each of the controller buttons are shown in Table 6-11.

Agent, you must put the controller into Action Control mode for these buttons to have any effect. In Action Control mode, you can't control your Spybot's movement, so be aware that you're more vulnerable to getting hit.

▼ **Table 6-11.**
Face Off Controller Settings

SETTING	DESCRIPTION
	The Force Forward button makes the opponent move in a forward direction. He is powerless to help himself.
	The Send Magnet button tells the opponent Spybot to move toward the nearest Spybot or controller for a while. You may use this button up to four times in a mission.
	The red button sends the Sedate, or Freeze, action to your opponent, which stops him dead in his tracks. You can use this button as often as you like.
	The Mind Control button is an interesting tactic you can use. It freezes your Spybot and then makes your opponent follow your controller commands. You can use this button twice in a mission, but remember to put your controller back into Remote Control mode!
	The Spinner action button causes the opposing Spybot to spin around for a while. You can use this defense any number of times.

When you have hypnotized your opponent, drive toward your home base. The other Spybot will follow you. Once you are close to your controller, you can "check in" with home base. Switch the controller to Action Control mode and place it near your Spybot. If you succeed, your Spybot does a happy dance.

This is a very tricky mission, Agent. If you have two Spybots, you might want to practice the mission by yourself against an enemy with no controller. Once you have mastered this, you can move on to competing against other agents.

Mission Settings

Table 6-12 contains the settings you'll use during this mission.

▼ **Table 6-12.**
Face Off Mission Settings

SETTING	DESCRIPTION
	Set the length of time that your Spybot is sedated by moving the slider to a value between 1 and 15. A longer sedation time makes it easier for your rival to hypnotize you.
	Set the length of time that your Spybot is hypnotized by moving the slider to a value between 5 and 30. The longer your Spybot is hypnotized, the easier it is for your rival to draw you to his base.

Gamma Overload: One Agent, One Spybot

The object of this mission is to save the reactor at the S.M.A.R.T. Space Propulsion Lab from a meltdown condition. During a routine systems test, an accident has damaged the outer wall of the reactor. The resulting gamma radiation leaks are causing the reactor to overheat, and there is not much time left to fix it!

Agent, your job is to locate the leaks using your Spybot's built-in Geiger counter and to seal them by creating a plasma shield.

> **NOTE** A *Geiger counter* is an instrument used to measure radioactivity. It makes a clicking noise that gets faster when it gets closer to a radiation source. Don't worry, Agent, when you hear the clicking sound from your Spybot, it's not measuring any actual radiation in your area.

Of course, there are other dangers waiting for you on this mission. Creating a plasma shield drains your Spybot's energy, so you will have to recharge it with the energy boosters from time to time. Also, exposure to radiation damages your Spybot control systems.

The designers of the reactor planned for this emergency. They have placed boosters around the reactor so that your Spybot can recharge itself when its energy levels get too low. If you find and all of the radiation leaks before time runs out, then you will succeed. Otherwise, you and your Spybot will have failed. Good luck, Agent.

Mission Setup

Here's what you'll need to do to set up for this mission:

- Clear an area of floor space. To increase the mission's complexity, add obstacles.

- Set up one or more similar desk lamps at the edge of the clear space. These are the energy booster stations.

- Establish a secure communications link with your controller and then switch the controller to Remote Control mode.

Mission Execution

Turn on your Spybot and wait for the middle red and green indicator lights to alternate. Place your Spybot under one of the lamps and press the Run button. You will hear a series of four rising tones that indicate the light sensor is calibrating. When the light sensor has finished calibrating, the yellow light will flash slowly and you will hear a steady series of tones.

 Place the Spybot facing toward the middle of the mission area and activate the touch sensor by moving the bumper. You will hear the mission start tones and all of the green indicator lights will turn on. The number of lit indicator lights tells you how much energy you have left.

Move your Spybot around the mission area using the remote control. Watch the yellow indicator light. If it blinks slowly, you are near a radiation leak! If you stop your Spybot, you will hear a ticking sound. Spin your Spybot around to look for the direction of the leak. As you get closer, the speed of the flashing and ticking will increase. When the flashing and ticking becomes very fast, you must fire the plasma beam to seal the leak.

The longer you take to find the leak, the greater the risk of damage to your Spybot's systems. The number of red indicator lights shows you how much damage you have taken. Your energy decreases rapidly as you fire the plasma shield, so remember to recharge your energy.

The booster points only restore energy—they do not reverse the radiation damage! Watch the green indicator lights to make sure you have fully recharged before you continue. Use your energy wisely and seal those leaks before the reactor overheats. The future of our space program is in your hands.

Mission Settings

Table 6-13 contains the settings you'll use during this mission.

▼ **Table 6-13.**
Gamma Overload Mission Settings

SETTING	DESCRIPTION
	Set the amount of energy used when creating a plasma shield by moving the slider to a value between 5 and 50. The more energy you use, the harder it is to complete the mission.
	Set the time required to recharge your Spybot's energy by moving the slider to a value between 5 and 50. Longer recharge times leave less time to find the leaks, which increases difficulty.
	Set the armor rating of your Spybot by moving the slider to a value between 0 and 20. Heavier armor makes the mission less difficult.
	Set the number of radiation leaks you must seal off to a value between 1 and 20 by moving the slider. The more leaks, the harder it will be for you to succeed.
	Set the total mission time in 30-second steps from 30 seconds to 10 minutes by clicking the plus (+) arrow or the minus (−) arrow. Longer missions are easier to complete.

Laser Maze: One Agent, One Spybot

The object of this mission is to escape from a factory protected by a network of invisible laser beams. Enemy agents have tricked your Spybot into looking for them in an old, broken-down factory. While you're busy looking for them in the wrong place, the enemy agents can raid a gold reserve.

Agent, your job is to escape from the factory and capture the thieves at the gold reserve.

Naturally, the bad guys are making things more difficult for your Spybot. They have installed a series of laser beams in the factory that will cause damage—and the beams are invisible to you! Your Spybot has special sensors that can detect the beams, and it will tell you which way to drive by using its indicator lights. It will even tell you if the beams have been turned off so you can make a dash for the door.

Hitting the laser beams will damage your Spybot's control systems, and any move you make will drain energy levels, so pay very close attention to those indicator lights! If you can navigate through the beams to the exit, you'll have a good chance to catch the thieves. If not, then you'll fail. Be careful, Agent.

Mission Setup

Here's what you'll need to do to set up for this mission:

- Clear an area of floor space. To increase the mission's complexity, add obstacles.

- Set up a desk lamp at the edge of the clear space. This is the factory exit.

- Establish a secure communications link with your controller and then switch the controller to Remote Control mode.

Mission Execution

Turn on your Spybot and wait for the middle red and green indicator lights to alternate. Place your Spybot under the lamp and press the Run button. You will hear a series of four rising tones that indicate the light sensor is calibrating. When the light sensor has finished calibrating, the yellow light will flash slowly and you will hear a steady series of tones.

 Place the Spybot at the opposite end of the mission area and activate the touch sensor by moving the bumper. You will hear the mission start tones and the red and green indicator lights will flash back and forth, indicating that the laser beams are inactive. You can start your dash for the exit.

Be careful, Agent—watch those indicator lights! They are the only way your Spybot has to show you where the laser beams are. You must hold the controller buttons for as long as the indicator lights are on steady if you want to avoid the laser beams.

When the lights indicate a spinning move, you must spin, not just turn. Table 6-14 shows the movements your Spybot partner requires you to make for the different indicator light states. If you need to find out how to spin the Spybot, refer to the "Handheld Controller/Beacon Modes" section in Chapter 3.

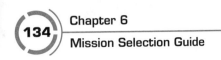
▼ **Table 6-14.**
Required Movements for Different Indicator Light States

INDICATOR STATE	DESCRIPTION
	When the two red indicator lights on the left side are on, spin to the left.
	When the two green indicator lights on the right side are on, spin to the right.
	When the leftmost red and rightmost green indicator lights are on, you must reverse the Spybot.
	When all of the indicator lights are on, stop and wait.

Remember that you must stop, spin, or go forward for as long as the indicator lights are on steady. Once they start flashing again in sequence, you can move any way you like. Use your energy wisely and get out of the factory without too much damage. S.M.A.R.T. is counting on you to get those gold thieves!

Mission Settings

Table 6-15 contains the settings you'll use during this mission. This is one of the few missions without a time limit.

▼ **Table 6-15.**

Laser Maze Mission Settings

SETTING	DESCRIPTION
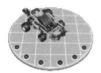	Set the complexity of the laser maze by choosing one of four levels. The higher the level number, the more complex the maze and the more points you can get.
	Set the strength of the laser beams by moving the slider to a value between 2 and 50. Stronger lasers will cause more damage and make it harder for you to succeed.
	Set the frequency of the laser scans by choosing one of five levels. Higher values mean the lasers will scan more often, and it will be harder for you to reach the factory exit door.
	Set the strength of your Spybot's armor by moving the slider to a value between 1 and 5. The heavier your armor, the easier the mission becomes but the fewer points you will get.

Robot Rescue: One Agent, Two Spybots

The object of this mission is to rescue a critically injured Spybot from a failed mission. The Spybot was sent in to investigate an energy leak, but it was left stranded when its own power ran out. You are the last hope to rescue the other Spybot from certain death.

You must get to the stranded agent and share your energy with her. This is similar to the technique of "buddy breathing" (see the following Note for more information) that stranded SCUBA divers use. Your energy levels will decrease, though, so be sure to return to the recharging station to keep your own levels up.

> **NOTE** *Buddy breathing* is when two SCUBA divers use one tank of air between them. If one diver runs out of air, the other diver can share his air supply so that both can make it back to the surface safely. It looks easy on TV, but just imagine what it feels like to be 60 ft. (20 m) below the surface with no air of your own.

When the other Spybot has enough energy, you must lead her back to home base and safety. Watch your own energy levels, Agent. If your run out of power before the end of the mission, S.M.A.R.T. will lose two of its best agents.

Mission Setup

Here's what you'll need to do to set up for this mission:

- Clear a large area of floor space. To make things more interesting, add some obstacles.

- Establish a secure communications link with each Spybot, then switch the stranded robot's controller to Action Control mode and put it on the floor at the far end of the mission area. This is home base.

● Switch your controller to Remote Control mode and put your Spybot near home base.

Mission Execution

Turn on the Spybots and wait for the middle red and green indicator lights to alternate. Press the Run button on each Spybot. When the Spybots are ready, the yellow lights will flash slowly and you will hear a steady series of tones.

To start the mission, place your Spybot near to the home base controller and activate its touch sensor. The indicator lights tell you how much energy you have. The yellow light on the stranded Spybot will blink.

Face the home base controller to boost up the energy reserves of your Spybot. Wait until all of the green lights are on before you try to rescue the stranded Spybot. Move your Spybot in front of the stranded one. When you are close enough, you can press the red button on your controller to transfer some of your energy.

Be cautious with your energy levels! As you transfer energy to the other Spybot, your own energy decreases. When you get back to home base you will have a chance to recharge, but make sure you have enough energy to get back.

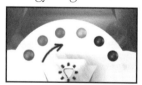

As the stranded Spybot gets more energy, it will automatically follow the rescue Spybot and try to get back to home base. Make sure that the home base controller is facing into the mission area and that you leave enough space for the other robot.

When both Spybots are in front of the home base controller, they will get a full charge and you will have completed your mission. Agent, I don't need to tell you to be careful with the red button, do I?

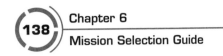
Mission Settings

Table 6-16 contains the settings you'll use during this mission. This is one of the few missions that does not have a time limit.

▼ **Table 6-16.**
Robot Rescue Mission Settings

SETTING	DESCRIPTION
	Set the amount of energy used by the Spybot motor by setting the slider to a value between 0 and 10. The more energy the motors use, the more you'll need to watch your energy levels.
	Set the amount of energy lost when the Spybot bumps into an obstacle by setting the slider to a value between 0 and 20. The more energy lost in a collision, the harder the mission becomes.
	Set the initial energy level of the rescue Spybot by setting the slider to a value between 40 and 100. Starting with less energy makes it more difficult to rescue the stranded robot.
	Set the amount of energy used per second by moving the slider to a value between 0 and 5. The higher the setting, the more energy your Spybot will use and the shorter the interval between recharges.

The Mole: One or Two Agents, Two Spybots

The object of this mission is to retrieve data files that have been stolen from the S.M.A.R.T. servers by a rival Spybot—a mole in the organization. Your Spybot has stopped the transfer of the files to a satellite uplink, but the enemy still has some of the data.

Your job, Agent, is to get back the rest of the data and put it back on the S.M.A.R.T. agency servers.

To have the best chance at retrieving the data from the mole, you'll need to get close to and in front of the enemy agent. After you get the data, you'll need some time to process it. Be careful, Agent—when you are busy processing data, the other agent can steal the data back, so move into a position that makes it harder for him!

Once you have the rest of the files, you must return to your satellite uplink to transfer the data back to S.M.A.R.T. Headquarters. Even during this phase of the operation, the enemy might be able to get some of the data, so be careful.

This is a long and difficult mission, Agent. Whether you are up against an automatic enemy or one controlled by another agent, you must move carefully to hold on to the precious files. If you fail, vital information will be lost to the enemy agents.

Mission Setup

Here's what you'll need to do to set up for this mission:

- Clear an area of floor space for this mission. Obstacles increase the difficulty, but the mission is hard enough without them.

- Establish a secure communications link between each Spybot and its controller and then switch your controller to Remote Control mode.

- If you are playing by yourself against an automatic rival, set its controller to Action Control mode and place it on the floor at one end of the room. This represents the rival's satellite uplink.

- If you are playing against a human-controlled rival, set its controller to Remote Control mode.

Mission Execution

This mission is ideal if you have two Spybots and want to play by yourself. In this case, the enemy agent should be uploaded from the Rival profile. It is a very good idea to practice against the automatic agent before you go head to head with a human-controlled Spybot.

If you are playing against a human-controlled mole, his instructions are exactly the same as yours. Just like in real-life espionage, the good guys and the bad guys can have the same tactics and the same objectives!

Turn on the Spybots and wait for the middle red and green indicator lights to alternate. Press the Run button on each Spybot. When the Spybots are ready, the yellow lights will flash slowly and you will hear a steady series of tones.

 To start the mission, activate the touch sensor on either of the Spybots. You will hear the mission start tones and both Spybots will show three red lights on, which means that each Spybot has half of the files and must try to get the rest.

Drive your Spybot around using the remote control. If one of the Spybots has the Rival personality, you'll see that it will try to get in front of you to start a file intercept. You must try to get in front of the enemy agent and then press the red button on your controller to start the transfer process. Be careful, your enemy will try to do the same thing.

When you have captured the data, you'll need some time to process it. The yellow light will blink to let you know that processing is taking place. It's a good idea to turn around and avoid the enemy to keep your data safe during processing. Also, once processing has started, there's no point in trying to get more data until the yellow light stops blinking.

As you capture more data, your green indicator lights come on. When all of the lights are on, you can attempt an uplink to the S.M.A.R.T. satellite. Race back to your controller, press the red button, switch your controller to Action Control mode, and place the controller on the floor in front of your Spybot.

During the uplink procedure, the enemy might still be able to get some of the files, so hurry back or you'll need to do it all over again! Remember to switch your controller back to Remote Control mode if you need to continue the mission.

This mission is all about positional strategy. Keep your Spybot moving to prevent the enemy from getting more of your files.

Mission Settings

Table 6-17 contains the settings you'll use during this mission. This is one of the few missions that does not have a time limit.

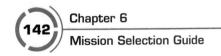
▼ **Table 6-17.**

The Mole Mission Settings

SETTING	DESCRIPTION
	Set the number of automatic Spybots in this mission from 0 to 1 by clicking the plus (+) arrow or the minus (–) arrow. The total number of Spybots must always be 2.
	Set the number of remote-controlled Spybots in this mission from 1 to 2 by clicking the plus (+) arrow or the minus (–) arrow. The total number of Spybots must always be 2.
	Set the time it takes to process data before an uplink moving the slider to a value between 0 and 20. Longer uplink-processing time gives the enemy more time to interfere and makes the mission harder.
	Set the number of files that must be uploaded to the satellite by choosing a value between 1 and 3. The number of files to send is actually twice the value that you set here. The more files you need to send, the harder the mission.
	Set the time it takes to process individual files after they have been captured from the enemy by moving the slider to a value between 2 and 10. Longer file-processing time means that the enemy has more time to interfere and makes the mission more difficult.

X-Factor: Two Agents, One Spybot

The object of this mission is to gather up as much X-Factor as you can before the mine elevator comes to take your Spybot out of the mine. X-Factor is the secret energy formula that powers all Spybots. It is collected from fire gems that can be found in the many passages of the deep mine.

There is a security system in the mine that is looking for intruders. If you are detected, the security system locks a fire beam on your Spybot and drains its energy.

You extract the X-Factor from the fire gem crystals by bumping into them with your Spybot. Each time you do this, your Spybot will need a few seconds to drain the energy from the crystal, so be careful to avoid the intrusion detection system.

You must be very wary, Agent, and keep as much energy as possible so that you can get back to the elevator when it arrives. The more X-Factor you can take to the elevator, the better. If you lose all of your energy to the intrusion detection system, you will fail.

Mission Setup

Here's what you'll need to do to set up for this mission:

- Clear an area of floor space and add a number of obstacles to represent the fire gems. This mission works best when you're in a fairly dark room.

- Set up your controller at one end of the mission area. It represents the mine exit.

- Get a flashlight. The other agent will use the flashlight to represent the security system. The other agent can point the flashlight at your Spybot.

- Establish a secure communications link with your controller and then switch the controller to Remote Control mode.

Mission Execution

Turn on your Spybot and wait for the middle red and green indicator lights to alternate. Place your Spybot under the flashlight beam and press the Run button. You will hear a series of four rising tones that indicate the light sensor is calibrating. When the light sensor has finished calibrating, the yellow light will flash slowly and you will hear a steady series of tones.

 Place the Spybot at the opposite end of the mission area and activate the touch sensor by moving the bumper. The indicator lights will all turn on and then they will go out one by one as a series of descending tones is played. Then all the lights turn back on to indicate your remaining energy. Now you can go back to the controller (this represents the mine exit) and start looking for the fire gems that hold the X-Factor.

 At this time, the other agent can try to track you down by trying to keep the Spybot's light detector in the beam of the flashlight. When the flashlight beam hits the Spybot light detector, you'll hear a high beep and all of the lights will flash. The other agent must keep the flashlight beam on the Spybot to draw out the energy. If the beam hits the Spybot for a long enough time, you will hear four low tones and then one indicator light will go out.

You can activate shields for about 5 seconds by pressing the red button on the controller. The Spybot will sound a rising tone. You can use these shields only six times in a mission. Keep in mind that each time you press the shield button, you reduce the number of shields by one and get 5 more seconds, even if the shields are already active.

When you hit a fire gem with the Spybot, the yellow indicator light will turn on and the Spybot will shudder briefly. Be careful, Agent, this is when you are in the greatest danger!

Near the end of the mission, the yellow indicator light will start to blink. This means that the mine elevator has arrived. Drive your Spybot toward your controller. When the Spybot is right in front of you, switch the controller to Action Control mode and place it directly in front of the Spybot.

This mission is all about keeping your Spybot on the move. Avoiding the tracking beam is your only hope of getting enough X-Factor before the elevator comes to take you to the surface.

Mission Settings

Table 6-18 contains the settings you'll use during this mission.

▼ **Table 6-18.**
X-Factor Mission Settings

SETTING	DESCRIPTION
	Enable or disable fatigue by clicking the on/off button. If you turn on fatigue, your Spybot will slow down as it loses energy, which makes it harder to win.
	Set the armor rating of your Spybot by moving the slider to a value between 1 and 6. Heavier armor makes the mission less difficult.
	Set the total mission time in 30-second steps, from 30 seconds to 3 minutes, by clicking the plus (+) arrow or the minus (–) arrow. Shorter missions are easier to complete.

Summary

This chapter has covered all the missions that come with the LEGO Spybotics CD. For more advanced missions and programming, see Chapter 7.

CHAPTER 7

MISSION CUSTOMIZATION GUIDE

U p to this point, you've learned all the important points that a S.M.A.R.T. agent needs to know to complete the preprogrammed missions. Now you'll take the next step and start creating your own custom missions that you can exchange with S.M.A.R.T. agents around the world.

This chapter covers the following topics:

- Examining the LEGO Spybotics Special Operations screens
- Choosing action capsules from the capsule magazines
- Adding functions to the remote control in Action Control mode
- Programming the Spybot to seek out a controller
- Designing a simple mission from scratch

Don't worry if you have no programming experience, Agent. The Spybotics Special Operations Center is designed to make learning fun and easy, and I'll present lots of tips that you can use to build your Spybotics expertise.

Introducing Spybotics Special Operations

The Special Operations screens are designed to make it easy for agents of all abilities to program their Spybots. You activate the Spybotics Special Operations Center (see Figure 7-1) by clicking the plus sign (+) icon on the communicator in the lower-left side of the home screen. The plus sign stands for additional functions.

Special Operations Button

▲ **Figure 7-1.**
Activating the Spybotics Special Operations main screen

Clicking the plus sign brings up the Special Operations Assignments screen (see Figure 7-2), which has a choice of missions for one Spybot, two or more Spybots, or even advanced free-form missions that you create from the ground up. Within each of these choices is a series of Special Operations Templates, or mission outlines.

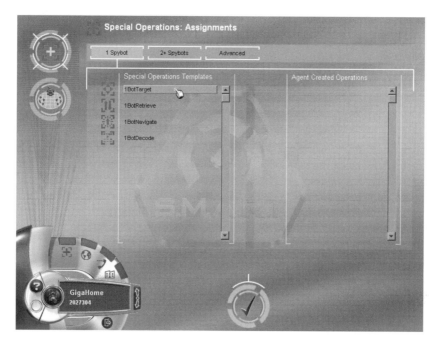

▲ **Figure 7-2.**
Spybotics Special Operations Assignments screen

TIP Back in the days before computer-aided drawing, drafting professionals used pencils and paper to create technical drawings. Often, they had to draw the same things again and again, so plastic *templates* were developed to make their job easier. A template would contain the outline for the shape the drafting professional had to draw. In Spybotics programming, you can think of the mission templates as basic outlines that you can fill in with details.

The basic templates available for different Spybot numbers are outlined in Table 7-1.

▼ **Table 7-1.**
Special Operation Mission Templates

ONE SPYBOT	TWO OR MORE SPYBOTS	ADVANCED MISSION
1BotTarget	2BotTarget	Free Agent
1BotRetrieve	2BotSurvive	
1BotNavigate	2BotPursue	
1BotDecode	2BotNavigate	

When you learn something new, it's best to know what it is you're expected to learn each step the way. In this chapter, you'll take each step of the Spybotics programming experience one at a time, and before you know it, you'll be designing your own missions. You must keep in mind the following three things as you discover the secrets of designing custom Spybot missions:

1. Start with the simplest ideas and move on only when everything is clear.

2. Confirm your understanding with simple experiments.

3. Pretend you are the Spybot when an operation is unclear.

With all of this in mind, let's start customizing a simple, single-Spybot mission. Click the 1BotTarget selection in the 1 Spybot mission template list on the Special Operations Assignments screen. Then click the check mark at the bottom of the screen to confirm your selection.

The Special Operations Logbook screen will appear (see Figure 7-3). In this screen, you have the chance to give your mission a name. The default name will be 1BotTarget1. You'll just be poking around here, so leave the default name for now. You can change it later. If you want to open a mission that you've saved, it will be in the list on the right side of the Special Operations Assignments screen.

The large blank area on the right of the logbook is for you to write up a brief description of your mission. It's always a good idea to fill in a bit of information here. I like to put the date and my S.M.A.R.T. username as the first items so that others will know who created the mission when they download it.

On the left side of every Special Operations mission template you'll see four buttons that you can use to navigate around your custom mission. Table 7-2 describes them briefly here. I describe them in more detail as needed later in this chapter.

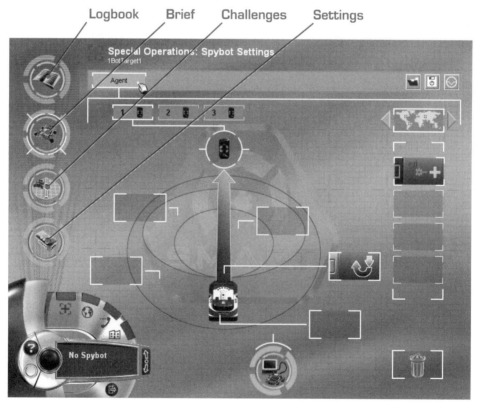

▲ Figure 7-3.

Spybotics Special Operations Logbook screen

▼ **Table 7-2.**
Special Operations Buttons

BUTTON	DESCRIPTION
Operation Logbook	Click this button to go to the main screen of each operation. Here is where you give your mission its name and write a brief description of the mission.
Operation Brief	Click this button to go to the Operation Brief screen. Just like the mission briefs in the predesigned missions, this is where you'll get the chance to see a video that explains the purpose of the mission. There is also some descriptive text on the right side of the screen.
Operation Challenges	Click this button to go to the Operation Challenges screen. Each of the mission templates has additional challenges that you can use to make the mission more or less complex. By clicking on the individual challenges across the top of the Operation Challenges screen, you'll get some ideas on how you can change the mission to make it more challenging.
Operation Settings	Click this button to go to the Operation Settings screen. This is where you can adjust the settings for specific parts of the mission. You can change things such as the mission time, the range of the laser, how often static disrupts your mission, and so on.

At the bottom of every Special Operations mission template is a Download Area button that gives you access to the Spybotics programming screens. Click the triangle-shaped button to bring up the Special Operations Download screen (see Figure 7-4). When a Spybot is connected to your computer, you'll see a picture of the Spybot at the end of the cable.

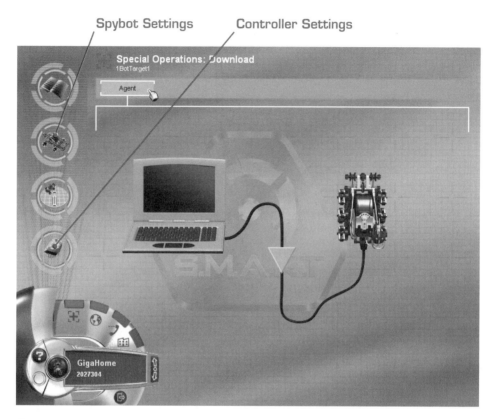

Spybot Settings   Controller Settings

▲ **Figure 7-4.**
Spybotics Special Operations Download screen

If you look closely along the left side of the screen, you'll see two new buttons. These are the Spybot Settings and Controller Settings buttons that I hinted at in Chapter 6. Table 7-3 describes these buttons.

▼ **Table 7-3.**
Spybot and Controller Setting Buttons

BUTTON	DESCRIPTION
Spybot Settings	Click this button to change the way your Spybot operates. For some types of missions, you might be able to actually run the mission automatically. For others, you will be able to give your Spybot special capabilities that can make the difference between mission success and failure.
Controller Settings	Click this button to go to the Controller Settings screen, where you can assign different functions to the controller buttons. If you want your Spybot to fire the lasers on command or spin to avoid detection, this is the screen you need to use.

You'll start learning about Spybotics programming by experimenting with the controller. You'll learn about the different kinds of function magazines and the action capsules within them. Finally, you'll download your experiment to your Spybot and make sure that everything is working.

Are you ready? Let's go, Agent!

Programming the Controller

The first step in designing custom missions is to understand how you can program the controller. In Chapter 3 you learned about the three modes that the handheld controller has: Link mode, Action Control mode, and Remote Control mode. Go back to Chapter 3 and review these modes if you are unclear about how they work.

Make your way to the Special Operations Download screen and click the Controller Settings button. The Special Operations Controller Settings screen (see Figure 7-5) will come up. In the middle of the screen is a picture of a controller with the selector switch in the Action Control mode position. This means that the action assigned to a particular button only works in Action Control mode.

TIP If you've been following along closely, you know that there's one button that works the same in both Action Control mode and Remote Control mode. That's right — it's the red button at the top of the controller.

Capsule Selector Arrows   Magazine

Capsule   Magazine Selector Arrows

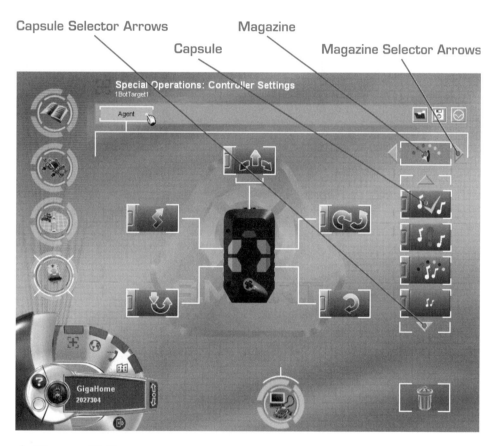

▲ **Figure 7-5.**
Special Operations Controller Settings screen

Magazines and Capsules

Along the right side of the Special Operations Controller Settings screen, you'll see a column of icons. Each column is called a *magazine*. Think of a magazine as a collection of similar types of actions. On either side of the magazine are arrows that spin the magazine rack to show you different groups or actions. These are called the *magazine selector arrows*. Within a magazine, there are *capsules* that you can move to the different holders on the controller. The magazine can show up to four capsules at a time. If more than four actions are available in a magazine, another set of arrows called *capsule selector arrows* will appear.

To move a capsule, click it and drag it over the holder that you want to put the action in. Click again to drop the capsule into the holder. If you move an action to a holder that already has a capsule, then the new action replaces the one that is already there. If you want to remove an action from a holder, just move it to the trash can in the lower-right corner of the screen.

> **TIP** If you want to know what any of the action capsules does, just move the mouse pointer over the capsule. The name and a brief description of what the action capsule does will appear until you move the pointer off the capsule.

When you first call up the Special Operations Controller Settings screen, the Mission Specific Capsules magazine is active. For the 1BotTarget mission, where the objective is to fire your laser, the only mission-specific capsule is the Laser Score action.

Assigning Action Capsules to the Controller

Click the magazine selector until the Special Effects magazine is active. Every mission, from the predesigned ones on the CD-ROM to a mission you create yourself, has the same actions available in the Special Effects magazine (see Table 7-4).

▼ **Table 7-4.**
Special Effects Magazine Capsule Contents

ICON	NAME	DESCRIPTION
	Special Effects Magazine	
	Whistle	Make a whistling sound.
	OK	Play the OK sound.
	Ouch	Play a warning sound and shudder from side to side.
	Hurrah	Play a victory sound and show a light sequence with the red and green lights.
	Whisper	Play a quiet sound.
	Hello	Play a "hello" sound.
	Done	Play a "task well done" sound.
	Morse	Play a Morse code. *
	Display Animation	Show a forward-pointing light sequence with all of the lights.

* *Morse code* is a way to transmit messages in which letters of the alphabet and numbers are represented by short and long signals. Back in the days before voice radio, people communicated across great distances using Morse code.

Let's see if you understand how this all works. In the next exercise, you'll assign a few of the Special Effects capsules to the controller.

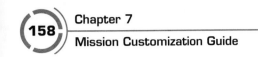

EXERCISE

Assigning Special Effects Capsules to the Controller

Move the capsules you want to assign to the buttons to their positions. When you're done, the screen should look like Figure 7-6. I've left the Laser Score capsule assigned to the red button.

TIP In some of the missions, the action capsules are locked in place if they are critical to the mission operation. You can tell when a capsule is locked because it has "clamps" above and below it in the holder. In the decoding missions, for example, the controller buttons are programmed to send the code digits to the Spybot, and you cannot change them.

To send the programming to the Spybot, click the Download Area button on the lower part of the screen, and then click the triangle to send the program to the Spybot. If you don't have a Spybot connected or if something goes wrong, don't worry. The online Trouble-shooter will help you get it working.

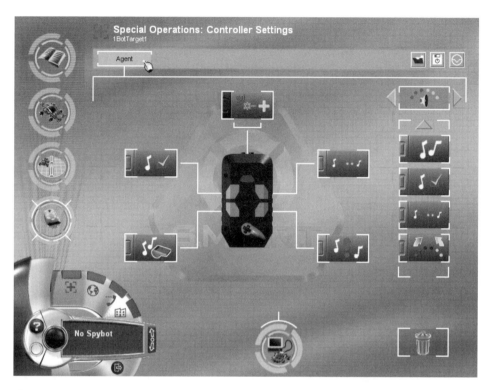

▲ **Figure 7-6.**
Controller settings for testing Special Effects capsules

Testing the Controller Settings

Now that you've downloaded the new settings to the Spybot, it's time to test them. Disconnect the Spybot from the cable and establish a secure communications link. As you're going to be doing a lot of testing, it's OK to leave the light sensor and laser light pipes disconnected for now.

Make sure that the controller is in Action Control mode and press one of the buttons. Nothing happens—what could be wrong? You haven't started the mission! The Spybot will always respond to the controller in Remote Control mode, but it only processes Action Control mode commands when a mission is running.

Just as with the standard missions, you start by pressing the gray Run button and waiting for the low tones that indicate your Spybot is ready to start the mission. Then you give the bumper a tap to activate the touch sensor. Now your Spybot will respond to the controller.

TIP If you programmed one of your controller buttons to activate the Morse Code capsule, can you figure out what the secret message is? A short tone is a "dot" and a long tone is a "dash." Press the controller button that plays the Morse code tone and listen carefully to the sequence. It's probably the most famous Morse code sequence ever transmitted.

The 1BotTarget mission is normally limited to 60 seconds. After this time, the mission ends, and if you have not met the default goal of scoring six laser target points, you fail the mission.

Experimenting with Movement Capsules

Before moving on, let's review what you've learned so far. You should fully understand the following:

- The different controller modes: Link, Remote Control, and Action Control
- How to navigate around the Special Operations Center
- How to navigate around the magazines and the action capsules in them
- How to assign action capsules to controller buttons
- How to download a mission to the Spybot
- How to start a mission on the Spybot

Now you're going to learn about the standard Spybot Movement capsules (see Table 7-5). Just like the Special Effects capsules, the Movement capsules are available in all types of missions. Their purpose is to make the Spybot move in certain ways depending on what is happening in the Spybot's environment. For example, if the Spybot hits an obstacle with its bumper, it can back up, turn a bit, and then begin moving forward again.

▼ **Table 7-5.**
Movement Magazine Capsule Contents

ICON	NAME	DESCRIPTION
	Movement Magazine	
	Search	Look left and right, and then go forward.
	Scan	Look left and right, and then wait.
	Spin Right	Spin to the right.
	Spin Left	Spin to the left.
	Explore	Spin right and then go forward to the left.
	Forward	Move forward.
	Backward	Move backward.
	Move Left	Spin left, go forward, and spin right.
	Move Right	Spin right, go forward, and spin left.
	Back Left	Move backward and then spin left.
	Back Right	Move backward and then spin right.
	Zig Zag	Move in a zigzag.
	Random Move	Move at random.

EXERCISE

Assigning Movement Capsules to the Controller

Once again you'll load up the controller buttons with some capsules, this time from the Movement magazine. By now, you're probably an expert at moving the capsules around. I've picked a few of the more interesting movements and placed them in the holders around the controller shown in Figure 7-7.

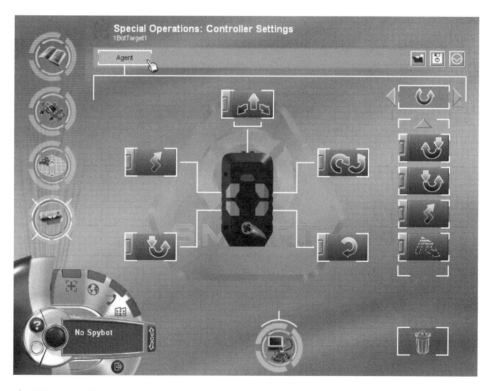

▲ **Figure 7-7.**
Controller settings for testing the Movement capsules

Download the program to the Spybot, establish a secure communications link, and start the mission. You can now test your Movement capsules by pressing the controller buttons. Notice that depending on the kind of Spybot you have, the movements will have slightly different ranges. For example, Shadowstrike will turn quite a bit faster and farther than Gigamesh.

That just about covers everything you can do with the Special Effects and Movement capsules on the controller. Remember, you can assign any action capsule to a controller button. This is an ideal way to test the operation of a capsule without having to deal with the additional complexity of how the Spybot is reacting to its environment—that's the next step.

Programming the Spybot

I've been dropping hints all through this book that the Spybot can run some missions all by itself. This is where you're going to learn how to program your Spybot to do this.

Start by restoring your Spybot controller settings to their defaults by clicking the Reset icon in the top-right corner of the Special Operations Controller Settings screen. Then click the Spybot Settings button to bring up the Special Operations Spybot Settings screen (see Figure 7-8).

The basic screen contents are the same for every mission. There is a top view of the Spybot, five holders for action capsules, and a target selection cell. When you program the controller, each holder corresponds to a specific button on the controller. When you program the Spybot, each holder corresponds to one of the Spybot's electronic systems.

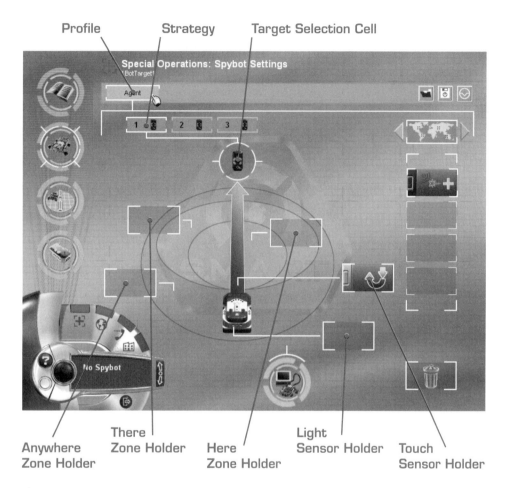

▲ **Figure 7-8.**
Special Operations Spybot Settings screen

If you recall the discussion on infrared range finding in Chapter 3, you know about the three ranges, or zones, that the Spybot can deal with: the Here zone, the There zone, and the Anywhere zone. You can put any action into the capsule holders that correspond to these zones.

There are two other capsule holders: one for the touch sensor at the front of the Spybot and one for the light sensor at the back of the Spybot. You can put any capsule into these holders as well. The capsule in the touch sensor holder is activated when the Spybot hits something with its bumper, and the action for the light sensor is activated when the Spybot sees a flash of light.

In front of the Spybot on the Special Operations Spybot Settings screen is an arrow pointing to a round cell. This is the Spybot Target Selection cell. You use this cell to choose from one of six possible targets that the Spybot can try to get to. Table 7-6 lists the targets and describes the icon in the cell.

▼ **Table 7-6.**
Spybot Target Selection Cell Contents

ICON	NAME	TARGET DESCRIPTION
	Black Outline	No target; only the Anywhere zone capsule is active.
	Black Controller	Select the nearest controller.
	Black Spybot	Select the nearest Spybot.
	Black Controller and Spybot	Select the nearest controller or Spybot.
	Green Controller	Select the nearest controller linked to this Spybot.
	Red Controller	Select the nearest controller not linked to this Spybot.

The final important feature that you'll need to be aware of is that the top of the Special Operations Spybot Settings screen may have more than one profile for each mission. For the mission you are working on there is only the Agent profile. Others, such as the Circuit Breaker mission, have profiles called Hacker and Jammer. Within each of these profiles, there can be one or more operational strategies. These strategies activate depending on what the Spybot is doing. For your 1BotTarget mission, the following strategies are available:

- Ready to fire at target
- Recovering from firing
- Spybot is experiencing interference from static

This is a lot of information for you to absorb all at once, but I have to put all the information in one place so you can refer back to it later. For now, you just need to know that a mission can have multiple strategies and that each strategy has places for you to put action capsules. Let's have a closer look at the strategies, Agent.

Decoding the Strategies in the Spybot Settings

Click each of the strategy boxes for the Agent profile and look at the default target call and capsule holder assignments. What do you notice? In each of the strategies, the default target is the black controller, and the Back Left action capsule is assigned to the touch sensor. What does this mean? Simply this:

For all strategies, target the nearest controller. Don't do anything if the target is in the Here, There, or Anywhere zones. Don't do anything if the light sensor sees a flash of light. Do the Back Left action if the touch sensor is activated.

How did I get from the pictures on the Special Operations Spybot Settings screen to the previous description? I used one of the most important tricks for understanding Spybotics programming: I pretended to be the Spybot!

EXERCISE

Pretending You Are a Spybot

Here's how to describe the actions of your Spybot in words, in five easy steps:

1. Start with the leftmost strategy box. Click it to show the target box and action holders for that strategy.

2. Describe the target cell.

3. Describe the action in the range-finding holders, starting with the Here zone, then the There zone, and finally the Anywhere zone.

4. Describe the action in the light sensor holder and the touch sensor holder.

5. If there are more strategy tabs, move to the next one and start the process again at step 2.

This seems like a lot of work, but once you get in the habit of pretending to be a Spybot, it is really quite simple. And the best part is that it works in reverse! If you can describe what it is your Spybot should do in each strategy, then you can put the correct action capsules into the holders.

Next, you'll try a little experiment to see if you understand how targeting works.

Moving Toward a Target

The first strategy available in the 1BotTarget mission is called *Ready to fire at target*. If you have a look at the Operation Settings, the target—in this case, the nearest controller—has to be in the Here zone for the Spybot to score a point. So how do you get the Spybot close to the controller? You need some special actions that move the Spybot depending on where the target is, and you can find those actions in the Target Based Movement magazine (see Table 7-7).

▼ **Table 7-7.**
Target Based Movement Magazine Capsule Contents

ICON	NAME	DESCRIPTION
	Target Based Movement Magazine	
	Point	Spin to keep facing the target.
	Point and Go Forward	Face the target and then go forward.
	Point and Go Backward	Face the target and then go backward.
	Point and Rest	Face the target and then wait.
	Advance	Move toward the target.
	Advance and Forward	Move toward the target and then go forward.
	Advance and Rest	Move toward the target and then wait.
	Retreat	Back away from the target.
	Retreat and Rest	Back away from the target and then wait.
	Turn Away	Turn away from the target.

These target-based movements are available in every single mission, just like the regular Movement and Special Effects magazines. They are slightly different from the normal movements. For example, whereas a regular Spin Left action will make the Spybot spin around for a while, the Point action makes the Spybot spin until it sees the target.

> **TIP** If you did the previous experiment with regular movements, you might have noticed that the Spin Left action or the Spin Right action sometimes turned the Spybot only a short distance before it stopped. This is because there were no action capsules in the Here and There target zones. If the controller is detected in the Here or There zone as the Spybot performs the spin action, the corresponding capsule is activated. If there is nothing in the capsule, then the Spybot simply stops spinning.

Learning about how each of the target-based movements works is something you can do on your own. Use the trick of assigning the actions to controller buttons that you learned in the previous section. Each Spybot works slightly differently due to its arrangement of wheels and gears, so it's a good idea to explore this in some detail.

Getting back to the task at hand, basically you need to move the Spybot close enough to the controller to score hits on the target controller. How can you do it? This is where *thinking like a Spybot* comes in handy. A typical problem-solving session might go like this:

Hmmm. I can't see a controller in the Here or There zone. That means it's probably on either side of my range-finding system. I'll just explore around a bit looking for the controller. When I see the controller in the There zone, I'll move closer. When the controller is in the Here zone, I'll fire the laser. If I hit something with the bumper, I'll just back up and spin a bit to the left—that way, I'll probably avoid hitting it again. Never mind the light sensor for now.

Hey—that's all the information you need to fill in the first strategy screen!

Filling In a Strategy Screen

Pick up the modules that do the actions you've just described and place them in the holders. When you're done, it should look something like Figure 7-9.

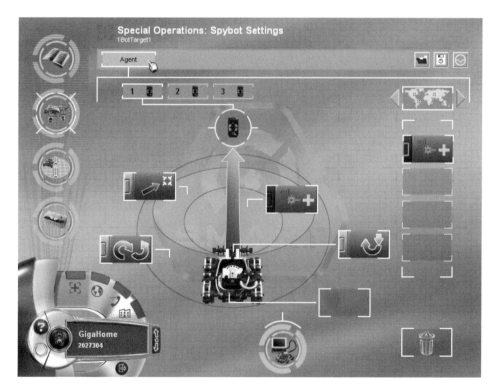

▲ **Figure 7-9.**
Filling in the *Ready to fire at target* strategy screen

Go ahead and try running the mission. Put the controller in Action Control mode and place it on the floor. Place the Spybot within about 6 ft. (2 m) of the controller and start the mission by pressing the Run button and then giving the bumper a tap. The Spybot should spin around and then advance toward the controller. When it gets close enough, the Spybot will fire its laser at the controller.

If you do this exercise a few times, you'll notice a couple of things. First, once the Spybot finds the target, it will just sit there blasting away until it gets six points or the mission time runs out. Second, sometimes the Spybot stops right at the edge of the Here and There zones, and the laser strikes have no effect!

You can fix this problem by replacing the Advance capsule in the There zone with the Advance and Forward capsule. Now the Spybot should move just a bit closer before it fires the laser, but sometimes it will still stop too far away.

OK, you've figured how to program the *Ready to fire at target* strategy. What about the other two strategies? What can you do to fill them in, Agent? Let's move on and find out.

Completing the Mission Strategy Screens

The two other strategies available in the 1BotTarget mission are *Recovering from firing* and *Spybot is experiencing interference from static*. If you click these strategy boxes, you'll see that all of the holders are empty except for the bump sensor, which has the Back Left capsule.

How you fill in the other holders depends on what the mission is all about. If this was a two Spybot mission and the other Spybot was trying to fire at you, then it would make sense to turn away from the target after firing your laser, right? The *Recovering from firing* strategy is activated after you fire the laser, and the best way to avoid the target would be to place the Turn Away capsule from the Target Based Movement magazine in the Here zone. You don't really need any actions for the other holders.

The *Spybot is experiencing interference from static* strategy is activated at random intervals based on your mission settings. In this case, no matter where the Spybot is, you probably want to simulate some kind of loss of control. The Ouch capsule from the Special Effects magazine is perfect for this. Place it in the Here, There, and Anywhere zone holders. Static can strike anytime, anyplace, after all!

> **TIP** Another tip to keep in mind is that the touch sensor and light sensor capsule holders can hold only one capsule for all of the strategies. That means that if the touch sensor is programmed with the Back Left action in the *Ready to fire at target* strategy, you can't use a different action in the *Recovering from firing* strategy.

When you have finished placing the capsules, upload your new settings to the Spybot and run the mission again. Notice the change in the Spybot's behavior after the laser is fired. You're well on your way now, Agent! You can probably handle programming a multi-Spybot mission yourself. Let's move on to the ultimate challenge: programming a Free Agent mission.

Programming a Free Agent Mission

Before moving on, let's review what you've learned so far. In addition to everything up to the section on Movement capsules, you should fully understand the following:

- How to read the contents of a Spybot programming strategy
- How to fill in a Spybot programming strategy
- How target-based movement can automatically control your Spybot
- How to combine strategies to accomplish a goal

This section will help you make the next step of designing your own mission from scratch. This is a pretty big step, though. Up to now, you've been given lots of guidance in terms of the names of the strategies and how the targets are set up. Free Agent missions remove almost all the guidance, so you're on your own.

> ▶ **TIP** Beginning programmers often make the mistake of tackling problems that are far beyond their abilities. In my consulting practice, I see many cases of programmers who have not completely mastered the basics of the problem they are trying to solve. Spybotics programming is no different. Hopefully, you've seen a pattern in the way I present the material here. Start with the simple stuff, understand it completely, and then move on.

Before you define your new mission, you'll need some background information on Operation Settings. You should refer back to this section again when you start designing your own missions.

Free Agent Operation Settings

Starting from the Special Operations Assignments screen, click the Advanced tab and open the Free Agent mission template. Fill in your Logbook with some basic information on this test mission. If you click the Operation Brief and Operation Challenges buttons, you'll see that the descriptions are pretty vague. You'll have to use a lot of imagination to come up with an interesting mission. Part of developing a mission is knowing what Operation Settings you have control over.

Click the Operation Settings button to review all of the things you can set up at the mission level. They are summarized in Table 7-8.

▼ **Table 7-8.**
Summary of Special Operations Operation Settings

SETTING	TYPE	DEFAULT	RANGE	DESCRIPTION
Mission Time	Arrows	0:00	0:00 to 10:00	Each arrow click adjusts the time by 30 seconds. A value of 0:00 means there is no time limit to the mission.
Points to Win	Slider	0	0 to 100	A value of 0 means that points don't matter.
Initial Points	Slider	0	0 to 50	The mission starts with some initial points.
Laser Power	Button	1	1 to 5	Higher values increase strength of the laser weapon.
Spinner Power	Button	3	1 to 5	Higher values increase power of the spinner weapon.
ElectroNet Power	Button	4	1 to 5	Higher values increase power of the ElectroNet weapon.

Continued

Continued

SETTING	TYPE	DEFAULT	RANGE	DESCRIPTION
Display	Button	Points	–	Depends on the mission. See Table 7-9 for more information.
Agents	Arrows	1	1 to 3	Sets the number of Agent profiles.
Rivals	Arrows	0	0 to 2	Sets the number of Rival profiles.

As you can see, there's a lot to keep in mind as you try to come up with a mission. You'll have some control over the weapon strength, the mission time, and even the kind of information shown on the display. In fact, the display settings are interesting enough to describe separately in Table 7-9.

▼ **Table 7-9.**
Summary of Special Operations Display Settings

DISPLAY SETTING	DESCRIPTION
Blank	The display capsules control what is shown on the display.
Time	The display shows fewer indicator lights as time runs out.
Points	The display shows more indicator lights as points are gained.
Radar	The display indicates the direction of the target.
Proximity	The display shows how far away the target is.
Random	The display lights blink at random.

That's a lot of information to keep straight, Agent! But don't worry—you'll start with a very simple mission.

Free Agent Mission Brief: Under Pressure

The 1BotTarget mission was very easy to program. All the Spybot had to do was find any target controller and then get close enough to it to fire the laser and score points. This mission will be a bit different. Instead of getting the Spybot to run the mission by itself, you'll take control of the Spybot for the first part of the mission and then you'll let the Spybot take control on the way home. You'll also use some obstacles to make the mission more challenging.

Here's the brief for the Under Pressure mission. You and your Spybot have entered a chemical plant that has been damaged in a fire. Inside the plant is a huge tank of dangerous gas that is about to burst due to the tremendous pressure from the heat of the fire. The only way to keep the gas contained is to flip over a pressure relief switch. But of course, you need to get a key for the switch, and the path to the key and the tank is blocked, so only your Spybot can get to it.

You need to move your Spybot to the locker where the key for the pressure relief switch is located, and then you need to get your Spybot to the tank to activate the pressure relief switch. Finally, your Spybot must get away from the tank as quickly as possible.

As you can see, the mission is divided into three phases that you must follow in order to achieve success:

1. Get to the locker where the switch key is stored.

2. Find the tank and activate the pressure relief switch.

3. Get away from the tank quickly (just in case).

Obviously, you need some way to run through the sequence to complete, and that's where two special action magazines available in a Free Agent operation that you can't find anywhere else come in handy. One is the Points magazine and the other is the Program magazine.

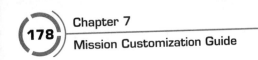

Free Agent Points Magazine

The Points magazine (see Table 7-10) determines what your Spybot must do to score points along the way. For example, if hitting an obstacle with its bumper should make the Spybot lose points, you would put the Penalty capsule into the touch sensor holder.

▼ **Table 7-10.**
Points Magazine Capsule Contents

ICON	NAME	DESCRIPTION
	Points Magazine	
	Give a Point	Give a point to another Spybot.
	Take Points	Take a point from another Spybot.
	Score Points	Score points.
	Penalty	Lose points.
	Reset	Reset score to 0.

Free Agent Program Magazine

The Program magazine (see Table 7-11) is the key to making the Free Agent operations work. The capsules in this magazine control the transition from one strategy to the next and can even determine whether you win or lose. For example, if you want to have a hit on the touch sensor signal mean that you win the mission, just drag the Win capsule to the touch sensor holder.

▼ **Table 7-11.**
Program Magazine Capsule Contents

ICON	NAME	DESCRIPTION
	Program Magazine	
	First Strategy	Stop the current strategy and start using the First Strategy.
	Second Strategy	Stop the current strategy and start using the Second Strategy.
	Third Strategy	Stop the current strategy and start using the Third Strategy.
	Fourth Strategy	Stop the current strategy and start using the Fourth Strategy.
	Fifth Strategy	Stop the current strategy and start using the Fifth Strategy.
	Lose	Play the lose sequence and end the mission.
	Win	Play the win sequence and end the mission.

Using the Program Capsules to Manage Strategies

Those Strategy capsules look like they might be useful, and in fact they are. By forcing your mission to flip to another strategy, you can make things move along in a particular sequence—and that's exactly what you need to do. Each of the three stages of the mission will be represented by its own strategy, and the last strategy will end the game when the Spybot gets back to home base safely.

Strategy 1: Retrieve the Key for the Switch

The first part of the operation is to get to the key for the pressure relief switch, which will be your First Strategy. You need to set up the holders in this strategy to define the mission, and then you need to make the Spybot move on to the Second Strategy after it gets the key.

Let's use a lamp to represent the pressure relief switch key. That means you'll need to put a Second Strategy capsule in the light sensor holder. When the Spybot detects a flash of light, it will move on to the Second Strategy.

You can make the mission more difficult by ensuring that if the touch sensor is activated, the Spybot drops the key and has to go back to the light source to get another one. Drag the First Strategy capsule from the Program magazine to the touch sensor holder.

This strategy has no automatic target, so there's no need to put anything in the target range holders. Set the target to a black outline to indicate no target. To let you know which strategy you're in, put the Animation capsule from the Special Effects magazine into the Here, There, and Anywhere zones. When you're done, the First Strategy screen should look like Figure 7-10.

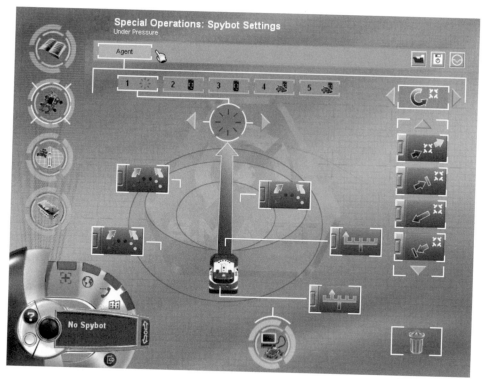

▲ Figure 7-10.
In Strategy 1, the Spybot retrieves the key for the switch.

Strategy 2: Flip the Pressure Relief Switch

The second part of the operation is to get to the tank, which is represented by the controller set to Action Control mode. In this strategy, you'll use the same target-based motion capsules as you used in the 1BotTarget mission template to move the Spybot closer to the controller. Instead of using the Laser Fire capsule when you get to the controller, put a Third Strategy capsule in the Here zone holder.

Remember that you can't change the light sensor or touch sensor actions in one strategy without changing them in all strategies, so hitting an obstacle will still cause the Spybot to go back to the First Strategy. This means that you'll need to go back to the light to get another key so you can continue the mission.

When you've finished setting up the Second Strategy, the screen should look like Figure 7-11.

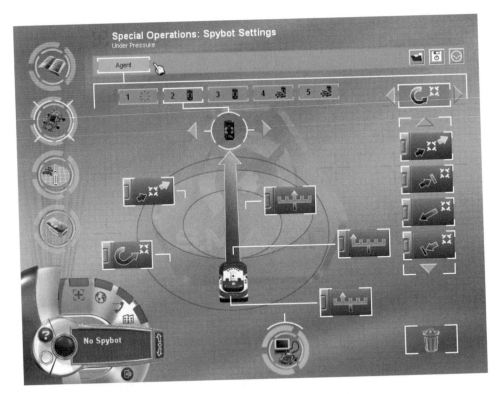

▲ **Figure 7-11.**
In Strategy 2, the Spybot finds the switch and flips it.

Strategy 3: Get Away from the Chemical Tank

The last part of the operation is to get as far away from the tank—or the controller—as possible. To do this, program the Spybot to back away from the controller while it's in the Here and There zones. In the Anywhere zone, place the Win capsule from the Program magazine.

If the touch sensor is activated before the Spybot gets to the Anywhere zone, you're back to the First Strategy and you must start all over again. If you manage to get to the Anywhere zone, the Spybot plays the theme song and you're done.

Figure 7-12 shows what the programming for the Third Strategy looks like.

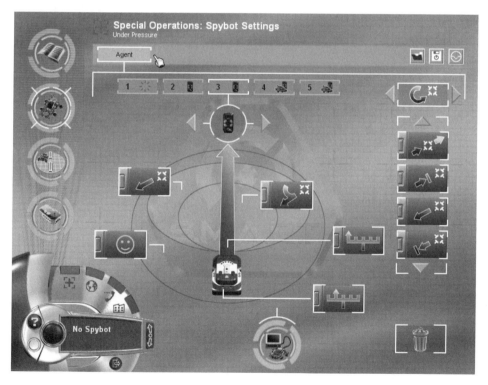

▲ **Figure 7-12.**
In Strategy 3, the Spybot backs away from the chemical tank.

Running the Under Pressure Operation

Running the Free Agent missions is just like running any of the prepro-grammed missions after you've downloaded the operation to the Spybot. You're using the light sensor in this mission, so calibrate the sensor by holding the Spybot under the lamp you're using to represent the key locker as you press the Run button. When the calibration is complete, the Spybot will make ticking sounds.

Make sure you have established a secure communications link with your Spybot and press the bumper to start the mission. Set the controller to Remote Control mode and move your Spybot to the lamp using the controller, but watch out for obstacles! You should see the indicator lights blinking.

After you get to the lamp, you're in the Second Strategy, so put the controller in Action Control mode and place it on the floor. The Spybot should try to find the controller and move toward it. Once it gets very close to the controller, the Third Strategy is activated.

In the Third Strategy, your Spybot backs away from the controller and then plays the winning animation. Note that the Spybot program-ming can be difficult to get just right. The boundaries between the Here and There zones can change as the batteries wear down. Also, if your Spybot detects a flash of light while it is backing away from the con-troller, it will jump to the Second Strategy, which moves it toward the controller.

All in all, you've covered a lot of ground here, so give yourself a pat on the back, Agent. You've come a long way!

Increasing Free Agent Mission Difficulty

If you want to have a real challenge, go back to the Operation Settings screen to make things difficult. Here are some options from the Operation Settings to think about:

Mission Time: By default, the mission time is set to 0:00, which means you have *unlimited* time to complete the task. Can you do it in 1 minute, or 30 seconds?

Weapons Power: If you programmed this mission with two Spybots, one of them could be trying to interfere with yours. If you change the weapons strength, you can make it more difficult for your Spybot to do its job.

Customizing Your Mission Area

One of the ways you can have even more fun with your Spybot partner and add more difficulty to the missions is to make your mission area more realistic. There are some easy things you can do to achieve this, such as

- Use full soda cans to make obstacles that are small but heavy enough to trigger your Spybot's bumper. You can design and color a pattern on a standard sheet of writing paper and wrap it around the can to give the obstacle a more finished look.

- Recycle cardboard boxes by cutting them up and gluing them to make ramps and barriers for your Spybots. The ramps can make certain missions such as X-Factor challenging by forcing the Spybot to get to the target controller along a narrow pathway. One wrong move and the mission is over.

- Draw the curtains closed to make the room a bit darker. The laser effect looks really cool in a dark room.

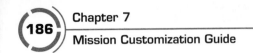

Summary

Congratulations, Agent! By programming this Free Agent mission, you have completed a very difficult part of your special training and you are now ready to design missions on your own. This will be a lot of fun, especially if you have more than one Spybot! Remember, you can always look at the preprogrammed missions if you need more hints about how to program the Spybots.

Appendix

MISSION-SPECIFIC CAPSULE CROSS-REFERENCE

In Chapter 7 you learned about the action capsules and the magazines that hold them. Some of the magazines hold capsules that are not available in all missions, so I've summarized all of them here.

Remember that you can add many of the action capsules to your LEGO Spybot for use in the preprogrammed missions that I summarized in Chapter 6. You can apply the ideas you develop in Special Operations programming to the other missions as well.

Part of the fun of LEGO Spybotics is discovering new strategies you can use to succeed in your mission. These tables contain valuable clues to the extra capabilities of your Spybot, such as which action capsules are assigned to which controller buttons in the missions.

Decoding the Tables

Don't be scared by the size of the tables in the next two sections. They have a lot of information, but it should be easy to figure out what you need. Every capsule that can be assigned to a Spybot or controller has a dot (•) in the column for the mission. Every capsule that is programmed to a controller button has the button number in the column for the mission. (For example, "1" means "button 1.")

Some of the missions have different profiles for multiple Spybots. The tables in this Appendix will make these capabilities clear to you, so study them carefully. Here is an example.

Let's say you want to find out if your Spybot has any special capabilities in the Face Off mission. Look along the top row of the Special Abilities table (Table A-1) and find the Face Off mission column. Within that column, you'll see headings for Agent Spybot, Agent 2 Spybot, Agent Controller, and Agent 2 Controller.

Look down the column for Agent Spybot. Every special ability you can program to your Spybot or controller has a dot (•) beside it. Now look down the column for Agent Controller. There are no button numbers filled in, so none of the Special Abilities capsules are programmed to the controller.

If you continue on to the Send to Target table (Table A-2), you'll see that button 1 (1) is assigned the Send Magnet action, and button 2 (2) is assigned the Send Mind Control action. Remember that you must be in Action Control mode for these buttons to work, but it might be very handy to use the Mind Control button to take over your opponent's Spybot!

Note that not all of the button assignments are shown in the Special Abilities table (Table A-1) and the Send To Target table (Table A-2). Some assignments are described in the Mission Specific Capsule summary (Table A-3). You'll have to read that section carefully to find the extra information I've included.

Special Abilities Capsules

Special Abilities are actions you can program into your Spybot or controller to give yourself special powers during a mission. For example, if you want your Spybot to be invisible for a while in the Face Off mission, just drag the Cloak capsule from the Special Abilities magazine to one of your controller buttons.

Check out Table A-1 closely, Agent. You might be surprised at some of your Spybot's hidden talents!

Send To Target Capsules

The Send To Target capsules (Table A-2) are part of the hidden arsenal of the expert Spybotics agent. You can use these capsules to control or even disable your opponent to give you more time to complete the task at hand. For example, if you want to slow your opponent down for a while in the Critical Countdown mission, drag the Send Quicksand capsule from the Send To Target to a button on your controller.

Be careful, your opponent might have access to this knowledge too!

Mission Specific Capsules

The Mission Specific capsules are actions that are usually available in only one mission at a time. Putting them in a big table didn't make much sense, so I've written a little bit about each capability and then outlined the missions you will find the capabilities in.

You might remember that you can assign action capsules to either the controller or the Spybot. In Table A-3, if an action also has a default assignment to a controller button, I'll specify the button number for each action.

▼ Table A-1.

Special Abilities Capsules by Mission and Profile Cross-Reference

Special Abilities

Special Abilities	Times per mission	Energy Crisis		Face Off				Circuit Breaker				The Mole				Critical Countdown				Laser Maze		X-Factor	
		Agent Spybot	Agent Controller	Agent Spybot	Agent 2 Spybot	Agent Controller	Agent 2 Controller	Jammer Spybot	Hacker Spybot	Jammer Controller	Hacker Controller	Agent Spybot	Rival Spybot	Agent Controller	Rival Controller	Agent Spybot	Rival Spybot	Agent Controller	Rival Controller	Guide Spybot	Agent Controller	Miner Spybot	Miner Controller
Quad Damage — Make the Laser, Spinner, and ElectroNet four times more powerful for a while.	3			•	•											•	•						
Shield — Protection against damage for a while.	6	•	2	•	•											•	•			•		•	2
Reflect — Return Laser, Spinner, and ElectroNet fire back to sender for a while.	2	•		•	•			•	•			•	•			•	•	1		•		•	
Cloak — Make your Spybot invisible to other Spybots for a while.	3	•		•	•			•	•			•	•			•	•			•		•	
Turbo — Set motors to the maximum speed for a while.	2	•		•	•			•	•			•	•			•	•			•		•	
Magnet — Move toward the nearest Spybot or controller for a while.	4	•		•	•			•	•			•	•			•	•			•		•	
Repulse — Back away from the nearest Spybot or controller for a while.	4	•		•	•			•	•			•	•			•	•			•		•	
Flash Blind — Ignore all Spybots and controllers for a while.	6	•		•	•			•	•			•	•			•	•			•		•	
Freeze — Stop moving and ignore controller commands for a while.	3	•		•	•			•	•			•	•			•	•			•		•	
Quicksand — Slow down motors for a while.	6	•		•	•			•	•			•	•			•	•			•		•	
Reverse — Reverse motor direction for a while.	2	•		•	•			•	•			•	•			•	•			•		•	

Robot Rescue	Rescuer Spybot	Stranded Spybot	Rescuer Controller	Stranded Controller	Command Override	Rogue Spybot	Rogue Controller	Gamma Overload	Geiger Spybot	Geiger Controller	1BotTarget	Agent Spybot	Agent Controller	1Bot Retrieve	Agent Spybot	Agent Controller	1BotNavigate	Agent Spybot	Agent Controller	1BotDecode	Rogue Spybot	Rogue Controller	2BotTarget	Agent Spybot	Rival Spybot	Agent Controller	Rival Controller	2BotSurvive	Agent Spybot	Rival Spybot	Agent Controller	Rival Controller	2BotPursue	Agent Spybot	Rival Spybot	Agent Controller	Rival Controller	2BotNavigate	Agent Spybot	Agent 1 Spybot	Agent Controller	Agent 1 Controller	Free Agent	Agent Spybot
												●												●	●				●	●				●	●				●	●				●
									●						●									●	●				●	●									●	●				●
	●	●				●			●			●			●			●			●			●	●				●	●				●	●				●	●				●
	●	●				●			●			●			●			●			●			●	●				●	●				●	●				●	●				●
	●	●				●			●			●			●			●			●			●	●				●	●				●	●				●	●				●
	●	●				●			●			●			●			●			●			●	●				●	●				●	●				●	●				●
	●	●				●			●			●			●			●			●			●	●				●	●				●	●				●	●				●
	●	●				●			●			●			●			●			●			●	●				●	●				●	●				●	●				●
	●	●				●			●			●			●			●			●			●	●				●	●				●	●				●	●				●
	●	●				●			●			●			●			●			●			●	●				●	●				●	●				●	●				●
	●	●				●			●			●			●			●			●			●	●				●	●				●	●				●	●				●

▼ Table A-2.
Send To Target Capsules by Mission and Profile Cross-Reference

Send To Target

Send To Target	Times per mission	Energy Crisis — Agent Spybot	Energy Crisis — Agent Controller	Face Off — Agent Spybot	Face Off — Agent 2 Spybot	Face Off — Agent Controller	Face Off — Agent 2 Controller	Circuit Breaker — Jammer Spybot	Circuit Breaker — Hacker Spybot	Circuit Breaker — Jammer Controller	Circuit Breaker — Hacker Controller	The Mole — Agent Spybot	The Mole — Rival Spybot	The Mole — Agent Controller	The Mole — Rival Controller	Critical Countdown — Agent Spybot	Critical Countdown — Rival Spybot	Critical Countdown — Agent Controller	Critical Countdown — Rival Controller	Laser Maze — Guide Spybot	Laser Maze — Agent Controller	X-Factor — Miner Spybot	X-Factor — Miner Controller
Fire Laser — Hit the target with the Spybot laser.		•						•	•			•	•							•		•	
Send Magnet — Force the target Spybot to move toward the nearest Spybot or controller for a while.	4			•	•	1	1									•	•						
Send Repulse — Force the target Spybot to back away from the nearest Spybot or controller for a while.	4			•	•											•	•						
Send Flash Blind — The target Spybot will be unable to see any Spybot or controller for a while.	6																						
Send Freeze — Force the target Spybot to stop moving and jam controller commands for a while.	3															•	•	5					
Send Quicksand — Force the target Spybot to slow down for a while.	6			•	•											•	•						
Send Reverse — Force the target Spybot to reverse all movements for a while.	2			•	•											•	•	4					
Send Mind Control — Take control of another Spybot for a while. Your controller is linked to the target Spybot and your own Spybot freezes.	2			•	•	2	2									•							

	Robot Rescue				Command Override		Gamma Overload		1BotTarget		1Bot Retrieve		1BotNavigate		1BotDecode		2BotTarget				2BotSurvive				2BotPursue				2BotNavigate			Free Agent
	Rescuer Spybot	Stranded Spybot	Rescuer Controller	Stranded Controller	Rogue Spybot	Rogue Controller	Geiger Spybot	Geiger Controller	Agent Spybot	Agent Controller	Agent Spybot	Agent Controller	Agent Spybot	Agent Controller	Rogue Spybot	Rogue Controller	Agent Spybot	Rival Spybot	Agent Controller	Rival Controller	Agent Spybot	Rival Spybot	Agent Controller	Rival Controller	Agent Spybot	Rival Spybot	Agent Controller	Rival Controller	Agent 1 Spybot	Agent Controller	Agent 1 Controller	Agent Spybot
	•	•			•		•	.			•		•		•				2				2		•	•				2	2	
																	•	•			•	•			•	•			•		•	•
																	•	•			•	•			•	•			•		•	•
																	•	•			•	•			•	•			•		•	•
																	•	•			•	•			•	•			•		•	•
																	•	•			•	•			•	•						•
																	•	•			•	•			•	•						•
																	•				•				•							•

Mission-Specific Capsule Cross-Reference

▼ **Table A-3.**
Mission Specific Capsules by Mission and Profile Cross-Reference

CAPSULE NAME/ CAPSULE ICON/ DESCRIPTION		MISSION	PROFILE	CONTROLLER BUTTON
First Aid	You can use this capsule three times in a mission to help repair damage to your Spybot.	Energy Crisis 2BotSurvive 2BotSurvive	Agent Spybot Agent Spybot Rival Spybot	— — —
Force Forward	Make the target go forward.	Face Off Face Off	Agent Spybot Agent 2 Spybot	4 4
Point Toward and Advance	Face the target and then move forward.	Face Off Face Off Robot Rescue	Agent Spybot Agent 2 Spybot Stranded Spybot	— — —
Special Fire ElectroNet	Make the target shake.	Face Off Face Off Critical Countdown 2BotTarget 2BotTarget 2BotSurvive 2BotSurvive 2BotNavigate 2BotNavigate Free Agent	Agent Spybot Agent 2 Spybot Agent Spybot Agent Spybot Rival Spybot Agent Spybot Rival Spybot Agent Spybot Agent 1 Spybot Agent Spybot	— — — — — — — — — —
Special Fire Laser	Hit the target with the Spybot laser.	Face Off Face Off Critical Countdown 2BotTarget 2BotTarget 2BotSurvive 2BotSurvive 2BotNavigate 2BotNavigate	Agent Spybot Agent 2 Spybot Agent Spybot Agent Spybot Rival Spybot Agent Spybot Rival Spybot Agent Spybot Agent 1 Spybot	— — 2 — — — — — —
Special Fire Spinner	Make the target spin around.	Face Off Face Off Critical Countdown 2BotTarget 2BotTarget 2BotSurvive 2BotSurvive 2BotNavigate 2BotNavigate Free Agent	Agent Spybot Agent 2 Spybot Agent Spybot Agent Spybot Rival Spybot Agent Spybot Rival Spybot Agent Spybot Agent 1 Spybot Agent Spybot	5 5 3 — — — — — — —

Continued

CAPSULE NAME/ CAPSULE ICON/ DESCRIPTION		MISSION	PROFILE	CONTROLLER BUTTON
Sedate	Immobilize the opponent so the other Spybot can be hypnotized by activating it with your Spybot's bumper. This action is locked onto the controller and cannot be removed.	Face Off Face Off	Agent Controller Agent 2 Controller	3 3
Point Wait	Face the target and then rest.	Circuit Breaker	Hacker Spybot	—
Random Bump	Play a warning sound, move backward a little, and then spin left some more.	Circuit Breaker	Hacker Spybot	—
Random Forward Spin	Go forward a little and then spin right.	Circuit Breaker	Hacker Spybot	—
Random Forward Spin Twice	Move forward slightly, spin right a little, go forward some more, and then spin left.	Circuit Breaker	Hacker Spybot	—
Jam	Jam the Hackerbot satellite realignment signal. This action is locked onto the controller and cannot be removed.	Circuit Breaker	Jammer Controller	2
Steal	Secure a file from the rival Spybot. This action is locked onto the controller and cannot be removed.	The Mole	Agent Controller	2
Avoid	Move forward, move a little to the left, then move forward, and finally move a little to the right.	Critical Countdown	Rival Spybot	—
Random Explore	Scan, turn right a little, go forward, spin left, and then go forward.	Critical Countdown	Rival Spybot	—
Retreat Here	Back away from the target and then turn and back away some more.	Critical Countdown	Rival Spybot	—
Retreat There	Turn left and then move backward if the target is to the left or move forward if the target is to the right.	Critical Countdown	Rival Spybot	—

Continued

Continued

CAPSULE NAME/ CAPSULE ICON/ DESCRIPTION	MISSION	PROFILE	CONTROLLER BUTTON
Scan Move Scan around, move forward, spin left, and then move backward.	Robot Rescue	Stranded Spybot	—
Send Energy Transfer energy from the rescue Robot to the stranded Spybot to help it back to base. This action is locked onto the controller and cannot be removed.	Robot Rescue	Rescuer Controller	2
Rest Forward Stop for a while and then move forward.	Command Override	Rogue Spybot	—
Rest Point Forward Stop for a while, and then face target and move forward.	Command Override	Rogue Spybot	—
Back Left Move backward and then spin left.	Command Override	Rogue Spybot	—
Back Right Move backward and then spin right.	Command Override	Rogue Spybot	—
Start Decode Start the yellow alert light blinking to signal that the rogue Spybot has started decoding the vault.	Command Override	Rogue Spybot	—
Send Code 1 Transmit 1 as the next decode guess. This action is locked onto the controller and cannot be removed.	Command Override 1BotDecode	Rogue Controller Rogue Controller	1 1
Send Code 2 Transmit 2 as the next decode guess. This action is locked onto the controller and cannot be removed.	Command Override 1BotDecode	Rogue Controller Rogue Controller	2 2
Send Code 3 Transmit 3 as the next decode guess. This action is locked onto the controller and cannot be removed.	Command Override 1BotDecode	Rogue Controller Rogue Controller	3 3

Continued

CAPSULE NAME/ CAPSULE ICON/ DESCRIPTION	MISSION	PROFILE	CONTROLLER BUTTON
Send Code 4 "4" Transmit 4 as the next decode guess. This action is locked onto the controller and cannot be removed.	Command Override 1BotDecode	Rogue Controller Rogue Controller	4 4
Send Code 5 "5" Transmit 5 as the next decode guess. This action is locked onto the controller and cannot be removed.	Command Override 1BotDecode	Rogue Controller Rogue Controller	5 5
Seal Leak Seal the gamma radiation leak. This action is locked onto the controller and cannot be removed.	Gamma Overload	Geiger Spybot	2
Laser Score Score hits by firing the laser.	1BotTarget Free Agent	Agent Controller Agent Spybot	2 2
Item Capture Get hold of some secret goods at the source.	1BotRetrieve	Agent Spybot	—
Item Delivery Deliver the secret goods at the goal for a complete retrieval.	1BotRetrieve	Agent Spybot	2
Auto Fire Point to the target, fire the laser, and then move backward to recover.	2BotTarget 2BotSurvive	Rival Spybot Rival Spybot	— —
Rest Cry Stop while the alarm sounds.	2BotTarget 2BotSurvive	Rival Spybot Rival Spybot	— —
Siren Retreat Back away from the target while sounding the siren.	2BotTarget 2BotSurvive	Rival Spybot Rival Spybot	— —
Laser Tag Pass the token to the current target by firing the laser. This action is locked onto the controller and cannot be removed.	2BotPursue	Agent Controller	2
Static Attack Move at random and then do a dance.	2BotNavigate 2BotNavigate	Agent Spybot Agent 1 Spybot	— —

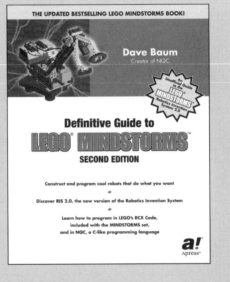
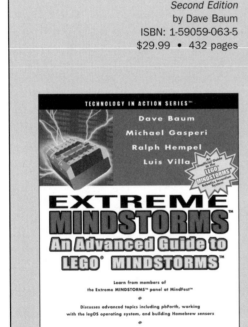

Apress Titles

ISBN	PRICE	AUTHOR	TITLE
1-893115-73-9	$34.95	Abbott	Voice Enabling Web Applications: VoiceXML and Beyond
1-59059-061-9	$34.95	Allen	Bug Patterns in Java
1-893115-01-1	$39.95	Appleman	Dan Appleman's Win32 API Puzzle Book and Tutorial for Visual Basic Programmers
1-893115-23-2	$29.95	Appleman	How Computer Programming Works
1-893115-97-6	$39.95	Appleman	Moving to VB .NET: Strategies, Concepts, and Code
1-59059-023-6	$39.95	Baker	Adobe Acrobat 5: The Professional User's Guide
1-59059-039-2	$49.95	Barnaby	Distributed .NET Programming in C#
1-59059-068-6	$49.95	Barnaby	Distributed .NET Programming in VB .NET
1-59059-063-5	$29.95	Baum	Dave Baum's Definitive Guide to LEGO MINDSTORMS, Second Edition
1-893115-84-4	$29.95	Baum/Gasperi/Hempel/Villa	Extreme MINDSTORMS: An Advanced Guide to LEGO MINDSTORMS
1-893115-82-8	$59.95	Ben-Gan/Moreau	Advanced Transact-SQL for SQL Server 2000
1-893115-91-7	$39.95	Birmingham/Perry	Software Development on a Leash
1-893115-48-8	$29.95	Bischof	The .NET Languages: A Quick Translation Guide
1-59059-041-4	$49.95	Bock	CIL Programming: Under the Hood™ of .NET
1-59059-053-8	$44.95	Bock/Stromquist/Fischer/Smith	.NET Security
1-893115-67-4	$49.95	Borge	Managing Enterprise Systems with the Windows Script Host
1-59059-019-8	$49.95	Cagle	SVG Programming: The Graphical Web
1-893115-28-3	$44.95	Challa/Laksberg	Essential Guide to Managed Extensions for C++
1-893115-39-9	$44.95	Chand	A Programmer's Guide to ADO.NET in C#
1-59059-034-1	$59.99	Chen	BizTalk Server 2002 Design and Implementation
1-59059-015-5	$39.95	Clark	An Introduction to Object Oriented Programming with Visual Basic .NET
1-893115-44-5	$29.95	Cook	Robot Building for Beginners
1-893115-99-2	$39.95	Cornell/Morrison	Programming VB .NET: A Guide for Experienced Programmers
1-893115-72-0	$39.95	Curtin	Developing Trust: Online Privacy and Security
1-59059-014-7	$44.95	Drol	Object-Oriented Macromedia Flash MX
1-59059-008-2	$29.95	Duncan	The Career Programmer: Guerilla Tactics for an Imperfect World
1-59059-057-0	$29.99	Farkas/Govier	Use Your PC to Build an Incredible Home Theater System
1-893115-71-2	$39.95	Ferguson	Mobile .NET
1-893115-90-9	$49.95	Finsel	The Handbook for Reluctant Database Administrators
1-893115-42-9	$44.95	Foo/Lee	XML Programming Using the Microsoft XML Parser
1-59059-024-4	$49.95	Fraser	Real World ASP.NET: Building a Content Management System
1-893115-55-0	$34.95	Frenz	Visual Basic and Visual Basic .NET for Scientists and Engineers
1-59059-038-4	$49.95	Gibbons	.NET Development for Java Programmers
1-893115-85-2	$34.95	Gilmore	A Programmer's Introduction to PHP 4.0

ISBN	PRICE	AUTHOR	TITLE
1-893115-36-4	$34.95	Goodwill	Apache Jakarta-Tomcat
1-893115-17-8	$59.95	Gross	A Programmer's Introduction to Windows DNA
1-893115-62-3	$39.95	Gunnerson	A Programmer's Introduction to C#, Second Edition
1-59059-030-9	$49.95	Habibi/Patterson/ Camerlengo	The Sun Certified Java Developer Exam with J2SE 1.4
1-893115-30-5	$49.95	Harkins/Reid	SQL: Access to SQL Server
1-59059-009-0	$49.95	Harris/Macdonald	Moving to ASP.NET: Web Development with VB .NET
1-59059-091-0	$24.99	Hempel	LEGO Spybotics Secret Agent Training Manual
1-59059-006-6	$39.95	Hetland	Practical Python
1-893115-10-0	$34.95	Holub	Taming Java Threads
1-893115-04-6	$34.95	Hyman/Vaddadi	Mike and Phani's Essential C++ Techniques
1-893115-96-8	$59.95	Jorelid	J2EE FrontEnd Technologies: A Programmer's Guide to Servlets, JavaServer Pages, and Enterprise JavaBeans
1-59059-029-5	$39.99	Kampa/Bell	Unix Storage Management
1-893115-49-6	$39.95	Kilburn	Palm Programming in Basic
1-893115-50-X	$34.95	Knudsen	Wireless Java: Developing with Java 2, Micro Edition
1-893115-79-8	$49.95	Kofler	Definitive Guide to Excel VBA
1-893115-57-7	$39.95	Kofler	MySQL
1-893115-87-9	$39.95	Kurata	Doing Web Development: Client-Side Techniques
1-893115-75-5	$44.95	Kurniawan	Internet Programming with Visual Basic
1-893115-38-0	$24.95	Lafler	Power AOL: A Survival Guide
1-59059-066-X	$39.95	Lafler	Power SAS: A Survival Guide
1-59059-049-X	$54.99	Lakshman	Oracle9i PL/SQL: A Developer's Guide
1-893115-46-1	$36.95	Lathrop	Linux in Small Business: A Practical User's Guide
1-59059-045-7	$49.95	MacDonald	User Interfaces in C#: Windows Forms and Custom Controls
1-893115-19-4	$49.95	Macdonald	Serious ADO: Universal Data Access with Visual Basic
1-59059-044-9	$49.95	MacDonald	User Interfaces in VB .NET: Windows Forms and Custom Controls
1-893115-06-2	$39.95	Marquis/Smith	A Visual Basic 6.0 Programmer's Toolkit
1-893115-22-4	$27.95	McCarter	David McCarter's VB Tips and Techniques
1-59059-040-6	$49.99	Mitchell/Allison	Real-World SQL-DMO for SQL Server
1-59059-021-X	$34.95	Moore	Karl Moore's Visual Basic .NET: The Tutorials
1-893115-27-5	$44.95	Morrill	Tuning and Customizing a Linux System
1-893115-76-3	$49.95	Morrison	C++ For VB Programmers
1-59059-003-1	$44.95	Nakhimovsky/Meyers	XML Programming: Web Applications and Web Services with JSP and ASP
1-893115-80-1	$39.95	Newmarch	A Programmer's Guide to Jini Technology
1-893115-58-5	$49.95	Oellermann	Architecting Web Services
1-59059-020-1	$44.95	Patzer	JSP Examples and Best Practices
1-893115-81-X	$39.95	Pike	SQL Server: Common Problems, Tested Solutions
1-59059-017-1	$34.95	Rainwater	Herding Cats: A Primer for Programmers Who Lead Programmers
1-59059-025-2	$49.95	Rammer	Advanced .NET Remoting (C# Edition)
1-59059-062-7	$49.95	Rammer	Advanced .NET Remoting in VB .NET